MINORITY MIGRANTS IN THE URBAN COMMUNITY

Minority Migrants
in the Urban Community

MEXICAN-AMERICAN AND NEGRO ADJUSTMENT
TO INDUSTRIAL SOCIETY

LYLE and MAGDALINE SHANNON

Department of Sociology and Anthropology
The University of Iowa

SAGE PUBLICATIONS · *Beverly Hills* · *London*

For information address:

SAGE PUBLICATIONS, INC.
275 South Beverly Drive
Beverly Hills, California 90212

SAGE PUBLICATIONS LTD
St George's House / 44 Hatton Garden
London E C 1

Printed in the United States of America

International Standard Book Number 0-8039-0158-5

Library of Congress Catalog Card No. 72-84055

FIRST PRINTING

ᒡ C

To the inmigrants of Racine whose labor has built a city
which is just beginning to recognize that
"the system has not equally rewarded all
according to their contribution"

FOREWORD

The original purpose of this project was to investigate the processes of value assimilation and behavioral change. Professor Robert McGinnis, formerly of the University of Wisconsin and now at Cornell University, and Professor Lyle W. Shannon, formerly at the University of Wisconsin and now at the University of Iowa, were co-directors of the project from 1958 through 1961. Professor Thomas J. Scheff, formerly of the University of Wisconsin and now at the University of California, Santa Barbara, and Professor Shannon were co-directors in 1961-1962. The latter has continued as principal investigator and director of the study since 1962.

Research was initiated with the support of the National Institutes of Health (Project RG 5342) and continued over a period of eight years with additional support from the National Institutes of Health (Projects RG 9980, GM 10919, CH 00042). Between 1958 and 1962 support was also provided at various times by the National Science Foundation, the Research Committee of the University of Wisconsin Graduate College, the Urban Research Committee of the University of Wisconsin, and the Ford Foundation grant to the University of Wisconsin.

Since 1962, analysis and writing have also been supported by the College of Liberal Arts, the Division of Extension and University Services, and the Graduate College of the University of Iowa. And more recently (1970) the Center for Studies of Metropolitan Problems of the National Institute of Mental Health has assisted us with a grant (MH 18196).

The community of Racine has always been most receptive to this project. It would be impossible to mention the many persons who have encouraged our research efforts and contributed in one manner or another. But from the very beginning two persons stand out as most helpful. The first is Professor Albert E. May, Director of the University of Wisconsin-Racine Extension Center at the time the interviews were conducted and now Professor of Mathematics, University of Wisconsin-Parkside. His efforts in providing research space and in opening doors throughout the community have been untiring. He continues to help us to this day. The second person is Mr. Leland Johnson, Director of Pupil Services, Racine Unified School District. It would not have been possible to select samples of the populations that we wished to study without his cooperation. He too must be thanked for his continued encouragement of our research and his untiring efforts to bring our findings to the attention of the community.

Judy McKim and Emily Meeks have assisted over a period of years with the data analysis and writing and rewriting of this volume; it could never have been finished without them. Credit for their contribution to the project has previously been given to over 90 graduate and undergraduate students, clerical and professional personnel in reports to the National Institute of Health and in papers in professional journals. Rather than list them again, let it be said that their work on the project at various times has not been forgotten. This final report has been made possible by the joint efforts of everyone from interviewer to the severest critics of the manuscripts. The faults that remain are ours.

PREFACE

The economic absorption and cultural integration of Mexican-Americans and Negroes into a predominantly Anglo host society are described in this volume. At the outset of the project considerable discussion took place between the researchers and representatives of each of these groups in order to decide upon a designation that would be appropriate for each, particularly the inmigrants. While people whose parents or grandparents have resided in either Texas or Mexico but are predominantly of Mexican descent may refer to themselves as Spanish Americans, Latin Americans, Hispanos, Chicanos, Mexican-Americans, or Mexicans, we desired to choose what they would like to have us utilize in reference to them. After considerable discussion with men and women of various ages the term "Mexican-Americans" was selected (1959); we use this term whenever we refer to the Racine sample. On the other hand, since we are comparing our sample with other populations or sub-populations, we must occasionally use the term "Spanish surname" or "white persons of Spanish surname" because this is the term employed in the 1960 U.S. Census. Specifically, when we wish to refer to the Mexican-American population for the United States we must be content with data on the Spanish surname population for the five southwestern states. Such confusion presents problems of comparability from time to time but we believe there is sufficient justification, nevertheless, for doing so because most of the Spanish surname people in the Southwest are what could be properly designated Mexican-American. Except in special reports, the white Spanish surname population is usually included with the white population.

A similar problem was presented in reference to the Negroes in 1960. At the time of the study most of the people with whom we were in contact wished to be referred to as "Negroes" rather than "blacks." We have continued to do so for the sake of consistency, although we are aware that "blacks" is now the preferred term. In addition, when we wish to make comparisons between our Racine sample of Negroes and other similar populations and sub-populations, the best information obtainable was listed as "nonwhite" population. Since nonwhites are predominantly Negro in the areas with which we were concerned, there really is no problem and the terms may be interchangeably used with no fear of misrepresentation.

The host society consisted of Anglos. They more frequently referred to themselves as Caucasians or whites but since Mexican-Americans were included in the study we decided (1959) that it would be best to designate the group as Anglos. When speaking with Negroes, we usually employed the term "white" for simplicity's sake—they would know about whom we were talking. Yet, in some parts of the Southwest Negroes are thought of as simply darker-skinned Anglos. The Census of 1960 referred most frequently to whites and nonwhites but in some reports presented data for Negroes.

Although we have attempted to be consistent from chapter to chapter and reference to reference, it is possible that on occasion we have referred to a statistic for the Negro population when we should have called it the nonwhite population, to a statistic for Mexican-Americans when it would have been more appropriate to use white persons of Spanish surname. We have attempted to minimize the references that might be misleading and believe that the reader will understand in spite of some difficulties in terminology.

Should the reader wonder why we have made so many references to Census data when it has, in a sense, created problems in comparison and exposition, let it be emphasized that we believe that the value of the study has been immensely increased by placing it in this national, regional, state, county, and community perspective. Since we have on other occasions commented on the shortcomings of studies that have either dealt only with communities of origin or with communities to which the inmigrants have come, or have not dealt with representative samples of the kinds of people about whom the researchers wish to make generalizations, it behooved us to make every effort to place our own data in proper perspective.

CONTENTS

Appendices

TABLES, MAPS, FIGURES

Chapter 1

THE SETTING

Focus of the Study

This book focuses on the adjustment problems of inmigrant Mexican-Americans and Negroes in a northern industrial community around 1960. "Inmigrant" is used to differentiate persons who are attempting to secure work and more or less permanently locate in the community from those who are in the migrant labor cycle and temporarily living and perhaps working in the agricultural areas nearby. We had defined our basic problem as one of discerning which factors facilitated or impeded the assimilation of new values and the modification of behavior among inmigrants who had left their predominantly rural places of origin, mainly Mississippi and Texas, and now dwelt in the community of Racine, Wisconsin. In our initial effort to understand how some persons were apparently more successful than others in adjusting to the urban, industrial society we were oriented toward the individual inmigrant and specifically concerned with his earlier statuses and his antecedent experiences as they related to his present characteristics. As the study progressed the findings suggested that the focus of attention should change from the individual to the social organization of the community—the most powerful determinant of the way of life of the inmigrant Mexican-Americans and Negroes.

As formal interviews and informal associations with Mexican-Americans and Negroes proceeded, it became increasingly apparent that what we were learning about the way in which the community was organized was telling us more about the range within which an over-all adjustment could be achieved by newly arrived persons than did their varied approaches to earning a livelihood and developing the kind of relationships with each other and the community that they desired. On looking back at the data, at the analyses so painstakingly conducted, and at the knowledge gained of how the social system in a northern industrial community operates, we could not help but conclude that a study more historical in perspective would have afforded a more complete description of the processes of adjustment. The study could have been more definitive had we paid greater attention to the week-by-week, month-by-month, and year-by-year fluctuations in employment as a consequence of the changing fortunes of various major heavy industries in the community. No less important, a more systematic investigation should have been made of both the formal and informal power structure of the community, not at the level most commonly studied by sociologists and political scientists, but at a lower level—the "little" power structure. The informal relations of individuals to each other (the nature of the interaction that takes place in the decision-making process determining who gets jobs and the kind of jobs upon entry into the community) should have had higher priority. Although a study of the gate-keepers in commerce and industry might not have been as exciting as studies of the larger power structure, attention to this group might well have paid unexpected dividends considering the fundamental concerns of the study.

Several strategies presented themselves as far as the basic research design was concerned and all were considered during the early period of the study before the interview schedule was constructed. It is interesting to note that the Mexican-Americans with whom we talked at the time seemed equally aware of these possibilities. The decision was to concentrate on measuring the extent to which each group has been absorbed into the industrial economy and integrated into the larger and extremely complex urban society rather than absorption and integration into either the urban Mexican-American or the Negro community, regardless of whether one or the other was a more propitious choice for inmigrant Mexican-Americans and Negroes. Furthermore, it was not our intent to produce a quantitative ethnography of two minority groups who had settled in a northern industrial community.

Though the main conceptual thought of this study turned toward the processes of economic absorption and cultural integration in a northern industrial community in the United States, the processual nature of its emphasis gives it general applicability throughout the United States and to other advanced or developing countries as well.

Rural-Urban Migration

During the nineteenth century there were intermittent movements of Mexicans across the southwestern United States border because of the threatened annexation of land to the United States, the gold rush, and railroad construction. In the early decades of the twentieth century the Mexican revolution, World War I, the curtailment of European immigrants, and the passage of the Oriental Exclusion Act of 1924 triggered a phenomenal increase in the number of Mexican laborers immigrating to that part of the country known as the Southwest. Industrialized farming, irrigation, and box car and home refrigeration were expanding the market for perishable fruit and vegetables grown in this area. The willingness of the unskilled Mexican laborer to accept "stoop labor" seemed a perfect and cheap solution to the labor problem.[1] In the second half of the 1920's Mexican immigrants accounted for more than 15 percent of all immigrants to the United States. After the depression years of the thirties during which many Mexicans were forced to emigrate, the Mexican flow across the United States border reached the same peak as before because of the "bracero" program—in eight of the eleven years from 1954 to 1964 more people came on immigrant visa from Mexico than from any other country. Although 1.3 million Mexicans legally entered this country between 1900 and 1964 for permanent residence, during the same period an undetermined number of "wetbacks" crossed the border illegally. Unfortunately, the term "bracero" labelled the Mexican-American as a foreign field hand who comes and goes, when in reality through succeeding generations most Mexican-Americans have been absorbed into the economy at higher levels. In the Southwest some joined the movement from rural farm laborer to non-farm occupations in the city. Others, roughly 400,000, entered the migratory labor cycle which took them to other sections of the United States following the seasonal harvest of grain, fruit, or vegetable crops and back to the Southwest for the winter. This reservoir of mobile labor existed because those in it had insufficient skill, education, or capital to do anything else.

> "When spring comes, whoever they are, they're the people who are the hungriest. Who else wants to work that hard for that little money?" (Personnel director of an industrial farm on the eastern seaboard.)

These workers were usually ineligible for relief, voteless, excluded from Old Age and Survivors pensions, without benefit of minimum wage laws, unemployment insurance, workmen's compensation laws, and the right to organize and bargain collectively. Tiring of such an existence, facing the threat of mechanization, and encouraged by word from more fortunate relatives in the urban centers of the North, many of them broke out of the "cycle" to try their luck in the city, to find a more permanent way of life for themselves and their children in less grueling occupations.[2]

Despite the apparent willingness of the Mexican immigrant to cross into the United States in increasing numbers, the chief source of rural reared migrants to cities in the United States has been from the rural areas *within* the United States. The urbanization of the American Negro, paralleling in many ways that of most rural migrants, has been distinctly different in other respects. Prior to 1860 most Negroes were concentrated in the South, a region of few large cities, and as slaves were mostly field laborers or in domestic service. The Civil War and emancipation were accompanied by great Negro mobility, but not by ready absorption into the economy. The slave status of the Negro, although legally absolved by the Emancipation Act, continued to impede his urbanization, particularly in the South. With the exception of domestic servants, a few skilled tradesmen, and some professionals, Negroes remained confined to farming and a few other non-industrial types of work.[3]

Substantial Negro rural to urban migration began around 1914; though the movement was first toward southern cities, World War I soon provided Negroes with greater opportunities for industrial labor positions than ever before in areas outside the South. Northern employers and letters from Negroes who had migrated to the North prior to this time stimulated migration toward the sources of increased demand for labor at the same time that the supply of Oriental immigrant labor was being drastically reduced. By 1920, 34 percent of the Negro population lived in cities compared to 19.8 percent in 1890.[4] Not only economic reasons but improved transportation facilities, restlessness, and the monotony and uncertainty of agricultural life, in contrast to the allure of the city, had served as stimulants for the migration.[5]

Charles S. Johnson in his book, *The Negro in American Civilization,* presents the changes that took place for a Negro as a consequence of his move to the North in the letter of a migrant to a friend back home:

"Dear M—:

I got my things all right the other day and they were in good condition. I am all fixed up now and living well. I certainly appreciate what you done for us and I will remember you in the near future.

"M—, old boy, I was promoted on the first of the month. I was made first assistant to the head carpenter when he is out of the place. I take everything in charge and was raised to $95 a month. You know I know my stuff.

"What's the news generally around H'berg? I should have been here 20 years ago. I just begin to feel like a man. It's a great pleasure in knowing you got some privilege. My children are going to the same school with the whites and I don't have to humble to no one. I have registered—will vote the next election and there ain't any 'Yes, sir'—it's all yes and no and Sam and Bill." [6]

World War II generated another wave of Negro migration to urban, industrial centers, not only to the North but to the western coastal areas as well. Between 1940 and 1950, 15 percent of the Negro population of the South left the southern states.[7] By 1960 the Negro population, with 73 percent in the cities, was more urban than the white, the latter being only 70 percent. Nine out of 10 Negroes in the North and West and six out of 10 Negroes in the South were urban residents.[8] "As of 1967 almost one-half of the total Negro population of the United States was living in the urban North and more than half of the southern Negro population was living in the growing urban industrial complexes of the South."[9]

Site of the Study

Founded in 1834 on the western side of Lake Michigan between Milwaukee and Chicago, Racine became well known for its manufacturing by 1887. Because of the prospect of employment and the easy all-water route to the new West, great waves of European immigrants were drawn to the port of Racine. Approximately one-third of its 1960 population of 98,144 was of Danish descent. Ninety churches, some of which are distinctly Polish, Armenian, Greek, or Scandinavian, attest to the diverse ethnicity of its residents.

In 1965, over 265 manufacturing firms employed approximately one-fifth of the population. Perhaps the best known of these industries are J. I. Case, S. C. Johnson and Sons (Johnson's Wax), and Western Printing. Although industry in Racine is notable for its production of tractors, farm implements, and wax products, it has been a producer of automobile equipment and accessories, electrical motors and appliances, leather goods, machine tools, electronic equipment, furniture, luggage, malted milk, lithography, apparel, food, and metal castings. The Chamber of Commerce described the city at the time of the study as follows:

> "The name of Racine is known to millions the world over because of one phrase: 'Made in Racine, Wisconsin, U.S.A.' For Racine is a city of industry. It is in industry that most people earn their living. And it is through industry that Racine has experienced her major historic success and now anticipates her greatest opportunity for future growth. . . .

> "for Racine was born as an incident in a great mass movement of people in pursuit of a dream . . . wanters and hopers on the hunt for the better life. . . . Throughout its history, Racine has been blessed with a plentiful supply of skilled, productive labor and this supply has continued to grow with the city."

> "Racine is a city of homes—a city of people with roots. Better than half of Racine's families own their own homes. That kind of a city attracts and holds industry. Racine is a city of craftsmen, mechanics, skilled workers, and executives. That kind of a city also attracts and holds industry."[10]

Although Mexican-American and Negro inmigrants did not receive the same welcome as that reserved for skilled persons for whom there is usually a need, many Mexican-Americans nevertheless came to the city because of promises of specific jobs or knowledge of specific places to look for work that had been transmitted to them by some friend or relative. Typical responses of Mexican-Americans were the following:

> "My brother-in-law arranged for my husband's job here in the factory."

> "We were a big family and people were always talking about how good it was here. My cousin wrote to me about jobs at —— —— and that I should come."[11]

The Situation in Racine: Impetus to Research

A considerable proportion of the Mexican-American inmigrants became permanent residents of Racine, though many returned to the Southwest for varying periods of time. By 1955, a community of approximately 200 households, mainly from Texas, had developed on the southern outskirts of Racine in an area known as Sheridan Woods. Despite the movement of many of its members to other parts of the city, this community continued to serve as a "reception center" for Mexican-American and Negro inmigrants, maintaining its size up to the time that the study began. Before the three-year span of field work was completed the processes of absorption and integration had changed the relatively homogeneous Mexican-American community of earlier years to a largely Negro one. Figure 1 shows the various school attendance centers of Racine according to the socioeconomic status of their residents in 1960 and Figure 2 shows the location of 800 persons interviewed at that time. Sheridan Woods is contained in the Hansche area.

Among the problems that provided impetus for this research were the poor physical health of the inmigrants and what seemed to be their economic dependence and instability. Communicable disease rates were high by public health standards and illnesses more common than among other groups in the community. The children of Sheridan Woods were considered to be ill-fed, ill-housed, and ill-clothed. There were neither indoor toilets nor running water in the community. The area was bordered on one side by a dump and on two other sides by the expanding city of Racine, which had refused to annex the area. In 1954 it was estimated that approximately 10 percent of the Mexican-American and Negro segments of the community were on regular public assistance roles; private welfare agencies dealt with a much greater percentage of the population. By Racine standards, the Mexican-Americans and Negroes who had recently arrived in the community had a low level of living.

During 1958 and 1959 before the introduction of formal interviews, frequent visits were made to the community in preparation for the formal survey to follow. Little or no difficulty was encountered in approaching those in the Anglo power structure of the community, representatives of its health and welfare agencies, or the leaders of the inmigrant minority groups. These informal interviews enabled us to determine how business and professional persons in Racine perceived Mexican-Americans and Negroes and how they in turn saw the community. What became rather apparent was that members of the host community tended to regard all Mexican-Americans and Negroes as a relatively homogeneous group regardless of the length of time they had lived in the community or differences in their educational backgrounds. The failure of persons in the larger community to discriminate between older and more recent inmigrants seemed upsetting to long-term residents among the Mexican-American and Negro groups. Some representatives of the various religious, medical, public health, and social service agencies as

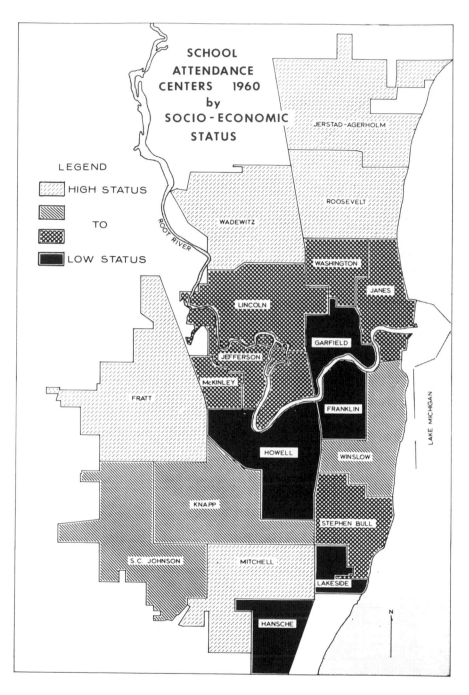

FIGURE 1.

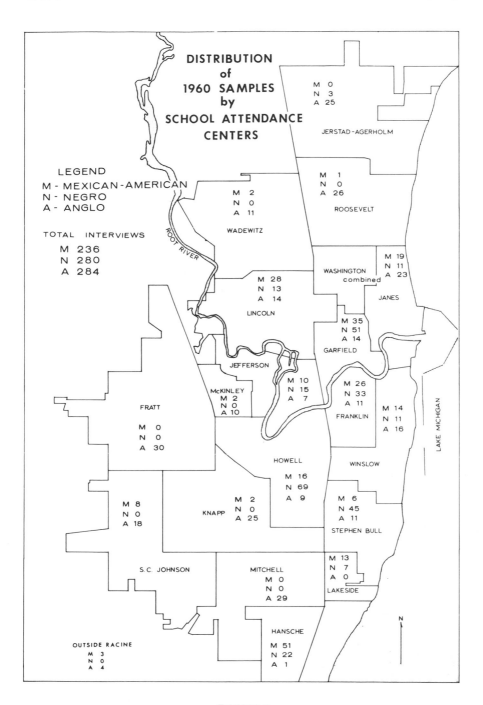

DISTRIBUTION
of
1960 SAMPLES
by
SCHOOL ATTENDANCE
CENTERS

LEGEND
M - MEXICAN-AMERICAN
N - NEGRO
A - ANGLO

TOTAL INTERVIEWS
M 236
N 280
A 284

ROOT RIVER

JERSTAD-AGERHOLM
M 0
N 3
A 25

ROOSEVELT
M 1
N 0
A 26

WADEWITZ
M 2
N 0
A 11

WASHINGTON combined
M 19
N 11
A 23

JANES

LINCOLN
M 28
N 13
A 14

GARFIELD
M 35
N 51
A 14

JEFFERSON

McKINLEY
M 2
N 0
A 10

M 10
N 15
A 7

FRANKLIN
M 26
N 33
A 11

M 14
N 11
A 16

LAKE MICHIGAN

FRATT
M 0
N 0
A 30

HOWELL
M 16
N 69
A 9

WINSLOW

KNAPP
M 2
N 0
A 25

STEPHEN BULL
M 6
N 45
A 11

M 8
N 0
A 18

S.C. JOHNSON

MITCHELL
M 0
N 0
A 29

LAKESIDE
M 13
N 7
A 0

HANSCHE
M 51
N 22
A 1

OUTSIDE RACINE
M 3
N 0
A 4

N

FIGURE 2.

well as other members of the community who by the nature of their work had a great deal of actual contact with inmigrant groups generalized about them in a manner indicating that their contacts were of a superficial nature.

Public health nurses, for example, maintained some contact with the inmigrant group; they gained entry into some households by checking all new babies listed by the hospital. If in their judgment there seemed to be a problem in the home that might affect the welfare of the infant they continued to return in an effort to make sure that it was receiving good care. In essence, the public health nurses seemed to regard residents of the Sheridan Woods community as a fine target population for extending standards of personal cleanliness so highly valued by them. In spite of their best efforts to overcome language barriers, they were extremely discouraged by their failure to effectively communicate with Mexican-American and Negro families. What they did not realize is that the problems confronted by change agents, in this case public health nurses, cannot be resolved simply by the process of communication. When housewives can be seen heating water for washing over a wood fire in a tub in the backyard, the number of baths and clean clothes possible for that family is measureably reduced. If the same house was equipped with bathtub, shower, hot and cold running water, soap, and towels, as well as washing machines, dryers, or any other facility taken for granted in the homes of the stable working class, the family would find it far easier to deal with the problem of cleanliness.

Some health and welfare personnel thought of all Mexican-American and Negro clients as chronic, incurable, and uncooperative. For example, when we asked agency personnel who were the "better people in the community," some of the names of the older Mexican-American and Negro families would be mentioned. Then, aware of the contradiction with their original superficial generalization they would retract their estimate of the inmigrants somewhat by saying, "Well, we only know the problem families, of course, so it is not fair to talk about all Mexicans as being this way."

Another assumption was that marital instability—desertion, or "coming and going," or coming in long enough to father a child and leaving—was more characteristic of the inmigrants than of the host population. Mexican-American husbands were also described as being heavy drinkers and gamblers. Some agency personnel believed that Mexican-American wives were poor housekeepers, those of Mexican origin a little better than those from Texas.

The more concrete issues of difference were housing and adequate nourishment for the family. Those in contact with inmigrant groups claimed that real estate agents tended to take the inmigrant only to the most blighted areas in the community and to persuade them to buy substandard housing at high prices. They were of the opinion that Mexican-Americans in particular had inadequate knowledge about diet and made themselves doubly susceptible to disease by being undernourished.

> "Nutrition is a big problem. The whole family is undernourished and this applies to the Negroes too, but it is even worse among the Mexicans. Mexicans tend to give the children soda instead of milk."

> "Pneumonia is a big problem among the Mexicans. They're not used to the weather and they don't take care of themselves. Since the babies are especially undernourished they get pneumonia."

Over-all, health and welfare personnel felt that the Mexican-Americans and Negroes were being exploited by various elements in the larger business community but that the inmigrants were likewise taking advantage of health and welfare personnel because they cooperated to the extent necessary to receive medicine and clothing but were unreceptive at other times and in other ways. Local doctors, though highly verbal about the unconforming way of life of the Negroes and Mexican-Americans, believed that with time in the community both populations would learn how to adequately take care of their children in spite of communication difficulties. What should have been realized was that health and welfare personnel, including local doctors, expected a responsiveness from the inmigrant typical of someone in the larger society. They were unprepared and dissatisfied with the reaction of persons who had come from another area and another subculture and who were unaware of the response expected of them. As a consequence, inmigrants with whom they had contact were often categorized as "undeserving" rather than "deserving" poor. Even if the inmigrants had been cognizant of what was expected in return for assistance, they were not capable of responding as fluently as expected by "middle class" professional persons.[12] Part of the difficulty in communication stems from language differences but an equal portion is related to differences in value systems.

Again and again comments received from persons in Racine indicated that they were reacting to what they defined as Negro or Mexican-American behavior, behavior equally characteristic of

Anglos in the lower socioeconomic categories. The reaction of school administrators to the drop-out rate and absence record of Mexican-American and Negro school children was a case in point. Since the general orientation of the school system in Racine was academic, rather than vocational, some school administrators perceived the system as less meaningful to Mexican-American and Negro children than to the typical Anglo student. One described the situation as follows:

> "Recently I had two families enroll seven children because the children had been in Arkansas working in the fields with the father. These children had missed a lot of school and one of the boys is quite a big boy and he is only in the fourth grade. In fact, I think he is 16, which means he can leave school at any time he wishes with only reaching the fourth grade. This amount of retardation seems to be common. In the City of Racine the compulsory school age is 18, but in the County of Racine and the State of Wisconsin, it is only 16. So it is quite common for the Mexican children not to have too much schooling."

Thus professional persons experienced great difficulties in their attempts to reach the Mexican-Americans and Negroes in the community. Health and welfare agencies, in particular, became aware of the fact that traditional social work approaches were not entirely adequate in dealing with what they defined as an extremely desperate situation. Yet they seemed to believe that their failures were based on the fact that members of the target group were relatively intractable rather than that their personnel and the traditional casework approaches were inappropriate and ineffective.

In the chapters that follow, the methodology and the findings of the study will be described in considerable detail. Several chapters will be devoted to a comparison of Mexican-Americans, Negroes, and Anglos in Racine with those who have remained in the communities from which the migrants came. Other chapters will concentrate on the adjustment problems of each group in Racine. In the next to the last chapter, the findings will be related to problems in the world of work, the educational institution, and public health and social welfare. And in conclusion, there will be a brief look at the respondents as they are today.

NOTES

1. The following volumes describe the historical position of Spanish-Americans and Mexican-Americans in the United States in great detail and in such a manner as to be particularly helpful in understanding the present position of the latter: Carey McWilliams, *Factories in the Field: The Story of Migratory Farm Labor in California;* Boston: Little, Brown & Co., 1939; Carey McWilliams, *North from Mexico: The Spanish-Speaking People of the United States,* New York: Greenwood Press, 1948 and 1968; Nancie L. Gonzalez, *The Spanish-Americans of New Mexico: A Heritage of Pride,* Albuquerque: The University of New Mexico Press, 1969.

2. The position of the Mexican-American during more recent years, particularly during the period of struggle for acceptance in the larger society, has been presented in what is probably the most definitive research report to ever be published on the subject: Leo Grebler, Joan W. Moore and Ralph C. Guzman, *The Mexican-American People: The Nation's Second Largest Minority,* New York: The Free Press, 1970. For a more popularized account of what has transpired in recent years see: Stan Steiner's *La Raza: The Mexican-Americans,* New York: Harper & Row, 1969 and 1970. A ginner's view of mechanization in the Southwest is presented in John McBride's *Vanishing Bracero: Valley Revolution,* San Antonio: The Naylor Co., 1963.

3. For an excellent comparative study of slavery in the United States and South American countries and the Negro's struggle for citizenship see: Frank Tannenbaum, *Slave and Citizen: The Negro in the Americas,* New York, Alfred A. Knopf, 1947.

4. *Fourteenth Census of the United States: 1920,* Volume II, Table 23, pp. 90-93; *Negro Population: 1790-1915,* U.S. Bureau of the Census, 1918, Table 5, page 90.

5. Morton Rubin portrays this picture of the migrant Negro in "Migration Patterns of Negroes from a Rural Northwestern Mississippi Community," Social Forces, Vol. 39, No. 1, October, 1960.

6. Charles S. Johnson, *The Negro in American Civilization,* New York: H. Holt & Co., 1930, p. 22.

7. "Population Estimates," Current Population Reports, Series P-25, No. 247, Table 4 (April 2, 1962).

8. "U.S. Summary," U.S. Census of Population: 1960, PC(1)-ID, Tables 158, 233. Census references will be abbreviated hereafter.

9. Herbert Hill, "Demographic Change and Racial Ghettos: The Crisis of American Cities," Journal of Urban Law, University of Detroit, Vol. 44, Winter, 1967, p. 232. For an excellent historical perspective on the development of Chicago's Negro ghetto see: Allan H. Spear, *Black Chicago: The Making of a Negro Ghetto, 1890-1920,* Chicago: The University of Chicago Press, 1967.

10. Descriptive Map of Racine, Wisconsin, Racine Chamber of Commerce, Racine, Wisconsin.

11. Many quotations found throughout this volume are from the actual interview schedules completed in the study. Interviewers were instructed to

reproduce as closely as possible word-for-word responses of the interviewee. Since Racine is continuously attracting inmigrants from rural areas in the South and Southwest as well as from the North, it is still an area in which the transformation of agricultural labor into industrial labor is easily observable. Thus, we have an industrial city to which persons with predominantly rural experience have moved from the South and Southwest, presumably in the hope of making a better life for themselves as did their forebears who founded and developed this industrial community.

12. The term "middle class" must be used with caution. The referent for middle class has never been completely agreed upon by sociologists. To some it refers to certain socioeconomic categories or socioeconomic status, and to others it refers to a life style or set of values. Here, in essence, it means persons from middle-range occupations, educational, and income backgrounds who are engaged in roughly similar occupations with roughly similar income and roughly similar educational requirements.

Chapter 2

THEORETICAL FRAMEWORK

Goals and Values and Changing Patterns of Living: Some Examples

Just as it is a natural phenomenon for man to be born, to live, to die, so too the transition from one way of life to another is at once a social process and a natural process. In the process of transition, for example, from a rural setting to an urban industrial milieu, people may experience a change in their values, life goals, and patterns of behavior. Through the process of value assimilation they may acquire, at least to a limited extent, the values and goals of the dominant group in their new home. It is, of course, neither necessary nor sufficient to acquire the goals and values of the larger community to bring about some transition in one's way of life. Though values and goals may or may not change, selected aspects of people's lives do change as a consequence of physical movement from one area to another, rural to urban, or southern urban to northern urban. Of an even more fundamental nature are changes in social and economic position that are almost certain to take place with movement from one place of residence to a distinctly different one. Should goals and values shift, resulting in higher (different) aspirations, changes in way of life (occupation, income, and level of living) may not take place in proportion to changing aspirations unless the community is organized in such a way as to permit it. Unless the social system is one in which alterations in goals and values may be translated into effective goal-achieving behavior, relatively little change in way of life or level of living may take place. The assumption that the problem should be approached within a social psychological rather than a social organizational or ecological framework is erroneous. Simple notions about the relationship of motivation to achieve and achievement may apply in a comparatively homogeneous community where everyone has relatively equal opportunity but they are not of much help in understanding what is going on in a complex heterogeneous northern industrial community such as Racine.

In order to sensitize the reader we have constructed from the data of the study a few typical types in different positions on the continuum of transition from one way of life to another.[1] José, Francisco, and Richard are Mexican-American inmigrants to Racine; each represents a different point on the scale from that of a recently arrived Mexican-American who has acquired few of the rewards available to those in the larger society to that of one who has relatively stable employment in the community and is beginning to acquire the end products of work in the larger society.

José

The interview with José, a comparatively new arrival from the Southwest, was conducted entirely in Spanish, the only language spoken in his home. His children were encouraged to speak English just during school hours. José resided with his wife and six children in a four-room apartment in a building which had been condemned. As the interviewer entered the home José expressed his concern over the eviction notice which had just come in the mail. Where would he find another home for his family at rent he could afford?

As the interviewer glanced about the apartment, he noticed the absence of a telephone, newspapers, and magazines. The furniture was very sparse and in poor condition. The family was seated around the television, the only visible source of entertainment. José was not much better or worse off than when he had been in the Southwest. There his home had lacked screens and a loosely hung curtain took the place of a door; there he had but two rooms and his family slept on the ricketiest of beds and cots. Because of the scarcity of work the family spent most of their time lounging around on the beds and the children played about the doorstep in the dust. One advantage, the thickness of the adobe walls, had made the place a bit cooler during the worst heat of the day, but there was no running water, which José did have in Racine. Other Mexican-Americans who had come to the northern community were not so fortunate and some did not even have the diversion of an operating television set.

José confided that he was worried about paying his bills and securing food for his family because he had been laid off for the past six months from his job of shoveling sand in the foundry. So far, he had managed with the assistance of the Racine County Welfare Department. He seemed very discouraged and, at one point, said that he sometimes wished he was doing agricultural work again for at least he would then be provided with a place in which to live. It was apparent that José and his family were barely getting along from day to day and that they had no real plans for the future. Although José occasionally liked to spend an evening at the tavern with friends, he could not do this very often; the only form of recreation for his wife was visiting with other Mexican-American women. When the question of schooling for his children was raised, José hoped that his boys would obtain some high school, although neither he nor his wife had gone beyond the fourth grade. He said, "You need learning to get a better job, like a mechanic." As for the girls, his wife could teach them most of what they needed to know and he would be proud of them if they got husbands and were as good cooks and housekeepers as their mother. "Well, they will always get married," seemed to satisfactorily summarize his responsibility toward them.

Francisco

The next interview took place in the home of Francisco, a less recent, better educated, inmigrant who put aside his evening newspaper to welcome the interviewer into his modest home. Actually, his family was quite comfortable and were pleased with the new, brightly colored linoleum in the living room which the landlord had just installed. Francisco indicated that he spoke English, but some questions necessitated rephrasing in Spanish before he could answer them. When Spanish was used exclusively toward the end of the interview he seemed quite relieved.

Francisco believed that his children had to go through high school. He believed that some college was necessary but doubted if he would be able to send his children as he could not possibly save for their education with the kind of money that he made, even with his recent promotion to assembly line inspector.

When asked about his leisure time activities Francisco showed the interviewer his LULAC membership card.[2] He was an officer of a local Mexican-American club and much of his social life centered around the meetings at a Mexican-American tavern. He had even thought of joining the P.T.A. since the local school principal had invited Mexican-American membership during the past year. He was hesitating until his English improved from talking with his Anglo friends.

Richard

The third interview was with Richard, who had been in Racine a few years longer than either José or Francisco. He spoke English very well and his responses to the questions were concise and clear. Just that day he and a friend had been discussing the Racine survey. He was eager to participate and had been wondering when the interviewer would come to his home.

Richard was employed at a level considerably higher than most Mexican-Americans in Racine, having a job as a bill collector for a local finance company. He was aware that his background aided him in his work since so many of the persons on whom he called were Mexican-American. In other ways it made the job more difficult, particularly when he knew he was better off than many of his compadres. He was able to pay his bills, was doing well, and was paying off the mortgage on his home. He was pleased that he was able to live in a relatively good school district since he valued an education highly. In fact, he had recently taken out an insurance policy which would enable him to start his boy in college. He confided to the interviewer that neither he nor his wife really felt close to the Mexican-American community in Racine, although they were Mexican-American. So many of their friends were Anglos that he was considered by them to be simply an American with a Mexican surname.

Although José, Francisco, and Richard have in certain respects the same antecedents, why is each at a different point on the continuum? Some people would contend that each is where he wants to be and others would say that Richard is the luckiest of the three. Our concern is with what transpires in the community which makes it possible for Richard to be where he is while José and Francisco have made fewer changes. How does the system operate so that some people move along more rapidly than others? José's life chances are not very good at this point, Francisco has a better chance, and Richard has it made.

Basic Concepts Employed in the Study

Acculturation

A general conceptual framework has been developed by sociologists and anthropologists who have dealt with processes of transition from one way of life to another. One of the basic concepts utilized by both disciplines, "acculturation," has value but it is doubtful if it has sufficient breadth to be of use as a guiding concept in this study. Acculturation, simply defined, involves the borrowing or acquisition of the cultural traits of one society by people from another society. The diet of José, who represents almost the beginning point on the transition continuum, is almost entirely Mexican but he buys bread in the supermarket or in the neighborhood grocery store, drinks Coca Cola, and has his wife prepare some Anglo food in the home. Richard, at the other end of the acculturation continuum from José, seldom has Mexican food and his wife frequently bakes Danish pastries. While José has a bright painting of a bull on black velvet hanging in his living room, Francisco has a picture of President Kennedy, and Richard has a print of a clipper ship above his bookcase of children's encyclopedias. But acculturation is a two-way process in which one acquires knowledge, beliefs, and behavior patterns similar to those in the larger host community of which one has become a part, and a process by which one also brings about some changes in the host society by introducing it to one's own culture. If you were to visit a Mexican-American restaurant in Racine you would find Anglos ordering tacos, tostados, and enchiladas. Being acculturated does not necessarily imply the abandonment of one's previous culture; it is possible to practice the old and the new simultaneously. The degree to which the acquisition of new behavior brings about the relinquishing of former behavior patterns varies with individuals and with the specific behavior pattern concerned.

Assimilation

Closely related to acculturation is the concept of "assimilation." The transfer of specific cultural traits may occur quite rapidly upon contact between people of different cultures, but the process of assimilation involves the gradual modification of social attitudes, the acquisition of new statuses and roles, and identification with new symbols. It is a process whereby the individual slowly comes to share the expectations of another group and slowly acquires a new set of definitions and values. Richard would be considered highly assimilated since he conceives of himself as belonging to the host community rather than to the Mexican-American community. The fact that Richard has been in Racine longer than José or Francisco probably helps to explain his greater assimilation into the host community, though it is probably only one of the factors responsible for the difference.

A distinction must also be made between the assimilation of individuals and that of groups. An individual may become invisible in various social contexts such as work, church, and leisure time activities but at the same time may maintain contact with a visible group of non-assimilated character through membership in an ethnic club or language society. The case of Francisco illustrates this point; although he may be invisible within the work situation, he is a member of a Mexican club which is highly visible within the host community.[3]

The use of the concept of assimilation now seems less appropriate than at the outset of the research since it is a longer-term type of process and would tend to focus our attention on the gradual processes by which one group becomes no longer distinct from another. Emphasis in this study is on the process whereby the inmigrant acquires the behavioral patterns of the larger society and learns how to play the major roles appropriate to his position in it.[4] Referring to this process as "absorption," S. N. Eisenstadt states, " . . . the process of absorption from the point of view of the individual immigrant's behavior, entails the learning of new roles, transformation of primary group values, and the extension of participation beyond the primary group in the main spheres of the social system."[5] Absorption can refer, of course, to the process of acquiring a position in any one, several, or all basic social institutions.

Economic Absorption

We concluded quite early in the study that "economic" absorption is such a powerful determinant of what happens in other spheres of life that major emphasis must be placed upon it. Economic absorption refers specifically to the processes of securing work and becoming a part of the regularly employed labor force in a given society. The decidedly broader concept of "cultural integration" refers to the process of integration into the whole gamut of institutional life. With this frame of reference we can say that the difference in the way of life of each of our three Mexican-Americans is related to the degree to which each has been absorbed into the economy. In other words, José has not been absorbed into the economy to the same extent as has Francisco or Richard.

Cultural Integration

Economic absorption may be thought of as a way-station to cultural integration. Though a limited amount of cultural integration, such as the ability to speak English, facilitates the inmigrant's entry into the economic institution and absorption into the economy, in turn, this absorption facilitates further integration into the culture of the host society. Once initiated, economic absorption usually proceeds faster than cultural integration because fewer changes in roles and conditional behavior are demanded.[6] The nature of the feedbacks in social processes has not always been taken into consideration by those who are concerned with this problem. But, as we have indicated before, economic absorption and cultural integration can take place to a limited extent without changing values. Changes in values can also take place without a parallel degree of economic absorption. For theoretical, empirical, and practical reasons (the reliability of economic data) we decided to emphasize the process of economic absorption rather than the broader concept of cultural integration.

As the study progressed it became increasingly clear that the degree to which inmigrants may be absorbed into the economy and integrated into the larger culture is determined to a major extent by the way in which the community is organized. How the larger social system functions is probably more decisive in determining what will happen to the inmigrant than how the inmigrant functions. At the same time, we cannot deny that the degree to which the inmigrant is accepted in the urban community is also related to his success or failure in assimilating the dominant values of the larger urban industrial society and in modifying his behavior accordingly. Thus, Richard is accepted in the larger community whereas José is not. Value assimilation is a specific type of assimilation and the understanding of this process is helpful in describing how inmigrants come to adopt the behavior patterns of the larger society.

Value Assimilation

Theories of value assimilation propose that values are acquired in the course of social interaction, either by direct experience or indirectly, through the mass media such as radio, TV, and newspapers. According to previous research, the crucial determinants of whether or not values will be assimilated are social psychological and sociological forces which lie outside rather than within the individual. (Whatever idiosyncratic [psychological] traits may intervene and influence the probability of value assimilation, they too have been acquired as a result of interaction with the surrounding social environment). Once assimilation of the values of the larger society has taken place among members of an inmigrant group, then changes in their personal behavior supposedly occur. In other words, it is presumed that value assimilation is an antecedent to significant behavioral change.

But value assimilation may take place without behavioral change and behavioral change may occur without value assimilation. Our contention has been that even though values have been assimilated, the social system or organization of society may make it difficult for the inmigrant to change some aspects of his behavior. For instance, the inmigrant's ties to the Mexican-American or Negro subcultural group may be so great that the cost of changing certain facets of his behavior to conform to the larger society would result in alienation from the one group in which he is at present accepted. The inmigrant knows that he is "different" but the larger society is organized in such a fashion as to make it difficult or perhaps even impossible for him to change. If, on the other hand, the inmigrant modifies his behavior in certain respects, not because he no longer "values" his social heritage but because by changing his behavior (perhaps his pattern of associations) to outwardly conform to that of the host society, it may help him get the job he wants or most probably needs. In such instances the inmigrant may perceive his change in behavior as still being consistent with his values. Behavioral change has taken place without value change. Accordingly, we should not have been surprised when analysis of the data revealed the correlation of values and behavior or sets of behaviors to be different among the Anglos than among the Mexican-Americans and the Negroes. It would have been most surprising had the results been otherwise.

Differential Association

When we began the study we hypothesized that the extent of value assimilation would be related to the nature of a person's associations within the larger society, that the behavior of inmigrants is related to the kind of associates and the nature of their relationship with them, both past and present. It is here that we found Sutherland's differential association theory very useful, a theory specifying facets of association relevant to the process of acquiring delinquent and criminal attitudes and engaging in subsequent behavior. In our application his theory has been broadened to say that those values will be assimilated which are first encountered, most frequently encountered, encountered for the longest period of time, and encountered in the most meaningful fashion. Accordingly, we were interested not only in what happens to people in the northern industrial community but likewise we were interested in the experiences they had in the community from which they came.

Sutherland's theory has been criticized as being entirely sociological, which is not really a fault but only a description of the nature of the theory, or as it might more properly be called, hypothesis. But to Sutherland's facets of association should be added several other dimensions: (1) the effect of social differences between persons or groups who are interacting in the process of assimilation; (2) the role of the person who initiates interaction; (3) the immediate consequences of interaction; (4) the inmigrant's perception of the interactional situation in terms of their past experiences. With this expanded set of the dimensions of value assimilation in mind we could justify asking most of the questions that were included in the interview schedule each year. The data were not always as rich as anticipated and as a consequence it turned out to be difficult to accept or reject all of the specific hypotheses that were generated by a theory including the dimensions that we considered necessary for an understanding of the process of value assimilation.

We hypothesized that value assimilation, cultural integration, and economic absorption would be related to various measures of association. The characteristics of inmigrants would differ from those of persons who had always lived in Racine as a consequence of their diverse antecedent associations and life experiences.

Each group would be likely to have values acquired from the members of groups in which interaction took place: (1) early in their lifetime, (2) frequently, (3) over a lengthy period of time, (4) in intensely personal situations, (5) where the social distance between interactors was such that the lower status person could conceive of himself in the position of the upper status person, (6) where the role of the interaction initiator was favorably defined by the lower status person, (7) where the consequences of past interaction were defined as favorable by the lower status person, and (8) where the immediate consequences of interaction were favorably perceived by the lower status person.

Mexican-Americans and Negroes, having spent most of their lives in a southern or southwestern subculture, should have a different set of values and behaviors than Mexican-Americans and Negroes who had always lived in Racine. Time in the community, to the extent that it permitted more associations of the nature that we indicated, would result in Mexican-Americans, Negroes, and Anglos becoming more similar. But if the community was organized in such a fashion as to minimize associations or to give associations a content that tended to rebuff the Mexican-American or Negro, the amount of value assimilation and behavioral change that would take place among them would be slight.

In summary, our initial efforts were spent in obtaining data on antecedent and current life experiences, hypothesized to be related to measures of cultural integration and economic absorption. As the research proceeded, measures of cultural integration were judged to be either inadequate or less reliable than were measures of economic absorption. Our emphasis turned more and more to the concept of economic absorption.

NOTES

1. The examples presented at this point are Mexican-American, although we could just as well have constructed Anglo examples as did W. Lloyd Warner and Paul S. Lunt in *The Social Life of a Modern Community: Yankee City Series, Volume One,* New Haven: Yale University Press, 1941. Although references will be made to representative Negro families at a later point, examples could well have been presented that would parallel the Mexican-Americans. It is assumed that most readers are familiar, however, with other sources in which Negro life styles have been portrayed, including inmigrants at different stages of absorption and integration. See, for example, E. Franklin Frazier's *Black Bourgeoisie: The Rise of a New Middle Class,* New York: The Free Press, 1957 and St. Claire Drake and Horace R. Cayton, *Black Metropolis: A Study of Negro Life in a Northern City,* New York: Harcourt, Brace & Co., 1945. Two more recent volumes describing varied Negro life styles developed in the process of coping with their urban environment are: Elliot Liebow, *Tally's Corner: A Study of Negro Streetcorner Men,* Boston: Little, Brown & Co., 1967 and David A. Schultz, *Coming Up Black: Patterns of Ghetto Socialization,* Englewood Cliffs: Prentice-Hall, 1969.

2. "LULAC or the League of United Latin American Citizens came into existence in 1929 as the result of efforts of leading Mexican-American organizations [The Order of the Sons of America, The Knights of America, and the League of Latin American Citizens] to join under one title and one set of objectives into a strong single instrument of unity. LULAC is an ethnical organization [sic] dedicated to securing and achieving dignity, respect, and unity

for themselves and their children in the eyes of their fellow Americans. They pledge themselves to the continuation of progress and social welfare and education for their ethnical entity [sic]." (LULAC News, Official Organ of the League of United Latin American Citizens, April 1965, XXVII, No. 4, in the foreword, "A Symbol of Dedication and Cultural Progress.")

There are, of course, more modern Mexican-American so-called "activist" groups such as the Alianza and the Brown Berets, who look down on the older LULAC organization as "hopelessly embroiled in and controlled by the Establishment and therefore incapable of improving the lot of those who claim to serve." They often refer to persons who cooperate with the Establishment as "Lulacks." Gonzalez, *The Spanish-Americans of New Mexico: A Heritage of Pride,* op. cit., pp. 180, 186-187.

3. For a thorough discussion of acculturation and assimilation see: Allen Richardson, "The Assimilation of British Immigrants in a Western Australian Community," Research Group for European Migration Problems, REMP Bulletin, Vol. 9, No. 1/2, January-June, 1961; and Wilfred D. Borrie, "The Cultural Integration of Immigrants," a study based upon the Papers and Proceedings of the UNESCO Conference on the Cultural Integration of Immigrants held in Havana, Cuba, April, 1955, Paris: UNESCO 1959.

4. Those who wish to consider the possibility of applying alternative sociological and anthropological approaches should consider: Frank E. Jones, "A Sociological Perspective on Immigrant Adjustment," Social Forces, Vol. 35, No. 1, October, 1956. The entire theoretical armament of sociology in reference to the adjustment of migrants is discussed.

5. See: S. N. Eisenstadt, *The Absorption of Immigrants,* New York: The Free Press, 1955, p. 15.

6. Borrie, op. cit., p. 101.

Chapter 3

A METHODOLOGICAL OVERVIEW

Preliminary Work in the Community

Continuous contact with members of the Mexican-American and the larger community began during the autumn of 1958. At that time the staff conducted a series of interviews with Anglos who were in positions of frequent contact with Mexican-Americans or who could refer us to persons with whom we should converse in order to develop a better understanding of the social organization of the Mexican-American community. The staff learned first-hand, so to speak, of the importance of the family, not only in the network of interrelationships that existed among people in Racine but also in the ties maintained by Mexican-Americans to their former places of residence. Simultaneously, the staff became acquainted with the Mexican-American population of the pretest community in Sterling, Illinois, the community from which we would obtain most of our Mexican-American interviewers. Small tape recorders permitted a more extensive record of their experiences during the day and lengthy discussions during the evening than would otherwise have been possible.

We considered extensive preparatory effort to be crucial to the success of the project, although such an effort to become well acquainted in the community is infrequently assigned as much importance by survey-research oriented sociologists. Some behavioral scientists would have been dismayed, considering the fact that the basic instrument of the study was to be a formal interview, to find that we were spending so many hours conducting informal interviews with Mexican-Americans and Anglos in decision-making positions in either social organizations of the larger community or the developing Mexican-American community. They would have completed the design of the interview schedule and begun interviewing in the field by the time that we were beginning to design the schedule. But, we contend, whatever we learned about Racine and Sterling before actually conducting pretests was still insufficient.

In summary, the goals of our preliminary work in Racine were: (1) to secure the cooperation of appropriate elements in both the Anglo community and the Mexican-American segment of the community, and (2) to inform ourselves about the social structure or social organization of the community, particularly as it related to determining alternatives available for mobility-oriented inmigrant Mexican-Americans. Knowledge of the latter would aid us in developing an appropriately structured interview schedule. In Sterling, a community where Mexican-Americans had lived longer and consequently were more fully integrated into the larger society, the emphasis of our preparatory efforts lay in: (1) securing Mexican-Americans who could be trained to become skillful interviewers for the work in Racine, (2) increasing our knowledge of the structure of a Mexican-American community in order to develop a meaningful interview schedule, and (3) pretesting the interview schedule in its various English and Spanish forms.

During the course of the autumn of 1958 and the spring of 1959 the staff conducted a series of unstructured interviews with representatives of public and private health, welfare, and educational agencies. In several instances the ensuing close relationship with agency representatives afforded access to official records touching on many of the problem situations of inmigrating Mexican-Americans. These accounts formed an important component of the larger study by providing a basis for comparison of responses to attitude questions.

Concurrently, personal visits were arranged with approximately 20 leaders living in those areas of the community in which Mexican-Americans resided in order to explain the study to them in detail and to ask them to play the role of "introducers." When a Mexican-American inmigrant seemed hesitant and about to refuse to be interviewed, the "introducer" would be called to approach him and vouch for the study. This procedure worked so well that out of a total of 256 Mexican-Americans in the 1959 sample only 21 direct refusals were encountered.

While conducting the 1959 survey, it became apparent that Negro inmigrants were facing essentially the same adjustment problems as were Mexican-Americans, albeit with some differences. The former were able to move into an older, well-established Negro community and they did not have the language handicap typical of a large proportion of the Mexican-Americans. To compare the two inmigrant groups in the process of adjusting to the host community would add an important dimension to the study; accordingly, it was decided to include a Negro sample in 1960. Contacts with leaders in the Negro community facilitated the interviewing process.

"Thank you" letters and attractive certificates of recognition were sent to the 1959 and 1960 respondents and to those who had assisted in the pretesting. The response of the community to our return in 1960 had been so favorable that there was little reason to believe that difficulty would be encountered when Negroes and Anglos were re-interviewed in 1961. There was none.

Basic Objectives of the Study

The basic objectives of the 1959 survey were: (1) to secure a description of fundamental values and interpersonal behavior among the Mexican-Americans and the Anglo controls, (2) to classify all respondents by socioeconomic status and to measure individual and intergenerational occupational mobility, (3) to gather data on migration patterns and work experiences of the Mexican-Americans and the Anglo controls, and (4) to test specific hypotheses that had been derived from value assimilation theory.

The basic objectives of the 1960 survey replicated those of 1959, but were more complex: (1) to obtain additional value and behavioral data for those persons in the 1959 Mexican-American sample who had children and to secure similar information for Mexican-Americans with children added to the 1960 sample, (2) to secure like data for a sample of the entire Anglo community with children, a sample which included the older, spatially contiguous, working class Anglos with children selected in 1959, (3) to obtain similar data on a sample of Negroes with children, (4) to improve the quality of the data on items with which interviewers and coders had difficulty in 1959, (5) to add crucial questions to the interview that were suggested by responses obtained in 1959, and (6) to secure data that would give the study a dynamic aspect by permitting the measurement of change over a period of one year.

Closely following the goals of the previous year, the 1961

objective was to secure: (1) value and behavioral data from samples of Negroes and Anglos interviewed in 1960, (2) more precise responses in areas of questioning that continued to be difficult (through modification in format), and (3) change data in order to maximize the dynamic aspects of the study.

The Interview Schedule

Designing the Interview Schedule

Originally, the interview schedule was developed to obtain data on four sets of variables believed to be crucial in understanding the basic processes of change in values and behavior. The first set consisted of the independent variables, sociological factors characterizing the former social environment of persons in each group, such as the nature of their place of origin, the occupational status of their parents, and the extent of their early urban-industrial exposure. The second set included intervening or mediating variables of a sociological and social psychological nature, characteristics believed to have been acquired or situations experienced during the interval elapsing between departure from the community of orientation and arrival in the current place of residence. These characteristics or experiences were presumed related to the independent variables but sufficiently uncorrelated to be treated separately from them, or truly separate and antecedent to the dependent variables, i.e., measures of absorption and integration. Examples would be job seeking and work experiences, and the attitudes, beliefs, or behavioral patterns developing from them. The third set of variables, the most difficult to measure to our satisfaction, involved values—the attitude and belief sectors of cultural integration. Among these were current attitudes toward mobility and world views. The fourth set were the dependent variables or end-behaviors hypothesized (by some sociologists) to follow from assimilation of the values of the larger society but which, as we have previously indicated, do not necessarily depend upon or follow from assimilation. These should be considered the behavioral aspects of cultural integration and economic absorption, as measured by occupation, income, and level of living.

We hypothesized that if behavioral change involves value assimilation as a prerequisite then behavioral aspects of cultural integration and economic absorption would be correlated with measures of values. If, on the other hand, value assimilation is not a prerequisite to behavioral change, then the intervening variables of a sociological (experiential) and a social psychological (attitudinal) nature would be more highly correlated with behavioral change than would the value assimilation variables. Specifically, level of education, job-seeking experiences, work experiences, experience in urban living, and patterns of association would relate more closely to measures of absorption and integration than would measures of values.

But, if it is hypothesized that the organization of the society into which the inmigrant is born is a more powerful determinant of level of absorption and integration at some later date than values (either the values acquired in community of orientation or values acquired in the urban-industrial milieu), then the earliest antecedent independent variables representative of experiences in the community of orientation should have a higher correlation with measures of cultural integration and economic absorption than either the intervening variables or current value indicators. And perhaps the combination of independent variables with intervening variables would produce even higher correlations with economic absorption and cultural integration. With these hypotheses in mind, the interview schedule was designed to obtain data permitting selection from these competing explanations of that which would best explain observed levels of economic absorption and cultural integration. None would be rejected with finality but low correlations between values and behaviors would raise serious questions about the importance of value assimilation as a prior condition to absorption and integration.

Pretests and Revisions

Designing and pretesting the schedule occupied the project staff from April to August 1959. The original version of the schedule was subjected to a total of nine revisions on a basis of pretest experience and discussion with a panel of Mexican-American interviewers. To avoid premature exposure of Racine's Mexican-American and Anglo population to the actual interview schedule, pretesting was conducted in Sterling, Illinois. Some of the members of three generations of interviewable Mexican-Americans in this city of 18,000 had been socialized in either Mexico or Sterling, while others had been socialized in the Southwest but had spent much of their life in the northern industrial community. Further revision of the interview schedule took place between the 1959, the 1960, and the 1961 surveys but the basic elements of the original schedule were preserved and the continuity of the study maintained. A special section on "black-white" relations which included the attitudes of Negroes toward interracial contact was added to the schedule commencing in 1960. Several new sections on social welfare agencies and other types of organizations concerned with the problems of the less fortunate were added in 1961 following conferences with members of the Mayor's Commission on Human Rights. The format of the schedule was also improved from year to year to facilitate the rapid and reliable recording of responses by interviewers.

Characteristics of the Samples Selected for Interviewing

A sample of 209 Mexican-American heads of households or their spouses, defined as ever-married, was interviewed in 1959. The sample was selected from a list compiled from the 1958 Racine City Directory and supplemented by extensive work in the community. The list included all Mexican-American families residing within the city limits of Racine and within the confines of Lakeside and Hansche school attendance centers directly south of the city limits. A control group of 189 Anglos, similarly defined, was also interviewed; it was selected from areas contiguous to the areas of Mexican-American concentration.

In 1960, stratified, systematic samples of 236 Mexican-Americans, 280 Negroes, and 284 Anglos were interviewed. The population from which these samples was drawn was narrowly defined as heads of households or their spouses with children 0-21 years of age. The Negro sample was entirely new. Those Mexican-Americans and Anglos who had been interviewed in 1959 and who had children were reinterviewed and formed part of the total 1960 samples. The Anglo sample was expanded so as to be more representative of the total Anglo population. In 1961, 137 Negroes and 189 Anglos from the previous year were reinterviewed. Sampling methodology is described in Appendix A.

A comparison of the samples according to occupation is presented in Table 1, by income in Table 2, and by education in Table 3. There were no significant differences from year to year within either the Mexican-American or the Negro samples. On the other hand, the income of the Negroes included in the 1960 sample was considerably higher than for those reinterviewed in 1961.[1] The Anglos in the 1959 sample had a lower occupational, income, and educational status than did those in the 1960 and 1961 samples; as previously stated, the 1959 sample came from areas contiguous to the

TABLE 1. PRESENT OCCUPATION OF HUSBAND: 1959, 1960, AND 1961 SAMPLES COMPARED

Occupational Status	Mexican-American		Anglo			Negro	
	1959	1960	1959	1960	1961	1960	1961
Professional, technical, managerial, proprietor	1	1	16	28	37	1	0
Clerical and sales	1	0	5	12	8	1	0
Craftsman, foreman	8	14	28	30	22	22	15
Operatives	30	26	18	16	15	33	36
Maintenance and service	1	1	9	4	8	7	4
Industrial laborer	45	46	11	7	7	22	29
Farmer and agricultural laborer	3	4	1	1	1	1	1
Non labor force, not ascertained, and inapplicable	12	7	12	2	2	13	15
	101	99	100	100	100	100	100

Mexican-American Samples: 1959-1960 χ^2 = 2.66, 3 d.f., not significant; Anglo Samples: 1959-1960 χ^2 = 17.98, 5 d.f., p $<$.01; 1960-61 χ^2 = 9.38, 5 d.f., not significant; Negro Samples: 1960-61 χ^2 = 5.91, 3 d.f., not significant; not ascertained and inapplicable responses eliminated in computing χ^2.

TABLE 2. TOTAL ANNUAL FAMILY INCOME: 1959, 1960, AND 1961 SAMPLES COMPARED

Income	Mexican-American		Anglo			Negro	
	1959	1960	1959	1960	1961	1960	1961
Under $3,000	11	12	20	4	5	12	20
$3,000–$3,999	9	13	8	5	3	9	12
$4,000–$4,999	32	28	16	8	5	16	9
$5,000–$5,999	19	18	16	16	15	17	17
$6,000–$6,999	10	11	13	20	23	17	17
$7,000–$7,999	4	5	5	12	16	10	9
$8,000–$8,999	1	3	3	11	11	3	3
$9,000+	2	3	7	16	18	1	3
Not ascertained and special circumstances	13	7	13	8	4	15	10
	101	100	101	100	100	100	100

Mexican-American Samples: 1959-1960 χ^2 = 4.59, 7 d.f., not significant; Anglo Samples: 1959-1960 χ^2 = 65.79 7 d.f., p $<$.001; 1960-1961 χ^2 = 4.86, 7 d.f., not significant; Negro Samples: 1960-1961 χ^2 = 8.13, 7 d.f., not significant.

TABLE 3. STATED EDUCATION OF MALE RESPONDENTS: 1959, 1960, AND 1961
SAMPLES COMPARED

Number of Years of Education	Mexican-American		Anglo			Negro	
	1959	1960	1959	1960	1961	1960	1961
00 Years	18	17	4	1	1	1	1
1—2 years	10	9	2	0	0	0	1
3—4 years	21	21	5	2	1	6	7
5—7 years	21	20	13	4	3	21	23
8 years	9	9	22	15	14	18	19
9—12 years	11	13	43	60	60	38	35
13+ years	2	1	7	17	20	3	3
Don't know, not ascertained, inapplicable	7	10	4	1	2	13	11
	99	100	100	100	100	100	100

Mexican-American Samples: 1959-1960 χ^2 = .18, 5 d.f., not significant; Anglo Samples: 1959-1960 χ^2 43.28, 4 d.f., p <.001; 1960-1961 χ^2 = 1.08, 4 d.f., not significant; Negro Samples: 1960-1961 χ^2 = .54, 4 d.f., not significant.

Mexican-American populations and thus represented a group of lower socioeconomic status. Characteristics of the Anglo sample for 1960 and 1961 more closely paralleled those of the larger Racine population.

The Interviewers

Most of the Mexican-Americans from Sterling who had assisted us during the pretest period were later employed as interviewers in Racine. They had been prepared by an intensive training program as well as by their experiences in the pretest situation. The Anglos selected for interviewing the control group were teachers and social workers; their education and previous interviewing experience made it possible to qualify them, it was believed, with a training program that was considerably less extensive than the one provided for the Mexican-American group.

In August of 1959, nine bilingual Mexican-American and nine Anglo interviewers entered the field. They interviewed from 9:00 A.M. to as late in the evening as respondents desired to schedule an interview. The greatest share of the interviewing took place over a two-week period with interviewers working six and seven days per week. During this time, interviews with 81 percent of the Mexican-American sample and 73 percent of the Anglo sample were completed; in other words, the proportion of unfinished interviews for Anglos was double that for Mexican-Americans.

The Mexican-American interviewers had a lower rate of refusal than did the Anglos. They had the use of introducers if a respondent refused or hesitated to give an interview and, having greater involvement in the project, they were more persistent and consequently more persuasive in securing interviews.

Since none of the previous year's interviewers were available in 1960, it was necessary to recruit and train a new group. In order to experience the lowest possible rate of refusal the training program was revised so as to involve prospective interviewers in as many facets of the project as possible over a long period of time.[2] Twenty-nine Negroes and Anglos, including some bilingual (Spanish-English speaking persons) took a 50-hour course including formal orientation sessions and lectures, interviewing techniques, coding practice, practice interviews, and actual interviewing experience. Fifteen were accepted, most of whom were highly motivated University of Wisconsin graduate students. Field procedures were similar to those utilized in 1959. The excellent rapport established with the previous year's respondents, the certificates of recognition which were sent to the previous year's respondents, and increased community knowledge and acceptance of the project resulted in fewer refusals than in 1959. In addition, the interview schedule was shorter and the Anglo interviewers were more persistent than their counterparts of the previous year.

In 1961, there were two groups of interviewers: those with experience from past years and a group of Anglo students who were participating in a National Science Foundation Undergraduate Research Participation Program. The latter was directed by the principal investigator of the larger project. Although the same field procedures were used as in previous years, this survey produced our highest rate of completed interviews, again primarily because of the persistence of the interviewers and their commitment to the project.

The Interviews

In 1959, the average interview lasted two hours; the 1960 and 1961 reinterviews were considerably shorter. Though interviewers and respondents were matched as closely as possible with regard to sex, race, and language each year, one of the main problems encountered during the course of the interviewing was gaining access to the home. The following suggestions were presented in the training manual given to prospective interviewers:

"The initial step in the interview is one which calls for strategy, firmness, and politeness. As to strategy you will be armed with a number of psychological advantages: (1) you will be a representative of the University of Wisconsin. This is a simple but effective label; it takes care of the possibility you might be selling something. (Your Identification Card will prove your point.); (2) you will have a clipping about the study cut out from the local paper. Not everyone will have read it, but it gives *you* the upper hand when you say, 'Perhaps you read about the survey in the paper' . . . and show the clipping; (3) you will be known to the authorities. Your name will be on file with the Racine Police Department, the local FBI office, and the Chamber of Commerce. It is very rare to be challenged, but it's best to be prepared; and (4) you will leave a 'Thank You' card with the respondent showing your name and referring the person with questions to Professor Davis at the Racine Extension phone number. All of these are your psychological weapons. BUT remember, you will not have to use them in most cases. *Do not* overwhelm the respondent with assurances that you are OK—act as if everyone takes it for granted. Use ID cards only if asked. You may flourish the clipping, but don't make a big point of it. *Don't* refer people to the police except in extreme situations. But *do* give the 'Thank You' card to the respondent. This normally would be at the end of the interview, but you may feel in some cases it would be best earlier. If you routinely give it at the beginning of the interview, the respondent will be distracted and fiddle with it throughout the interview."

Once access was gained the ideal type of interview was one in which the respondent's speech was fluent, the attitude was cooperative, and the interview was quiet with no other persons present. While a home atmosphere was probably most conducive to good interviews, in all cases the setting for Anglos tended to be more ideal than that for the Negro or Mexican-American interviews. This should not be interpreted as a function of membership in a race or ethnic group but as situational and stemming from socioeconomic differences. The following examples of responses to the question, "How did it go?" illustrate the point:

"Poorly, R. very suspicious. It took 15 minutes of talking to even get into the house. He frequently interrupted to inquire as to why I was asking so many questions and who was going to find out about him." . . . "R. is very race conscious. Very concerned over prejudice. Very concerned over talking to such a 'highly educated person.' He constantly stated that Mexicans were just as good as anybody." (Respondent was Mexican-American)

"Good, respondent was very cooperative but he spoke softly and slowly.

He had difficulty understanding some questions but after he understood he was very cooperative. At first it was quiet but then the children and the ladies came in and it was almost impossible to keep going." (Respondent was Negro)

"Rotten. This fellow had no ideas or opinions about anything. I would ask him a question and he would say 'more' or 'same' and had no comments at all to make except 'I don't know.' We had a lot of language trouble. The Spanish he spoke was poor, and one of his boys helped to translate once in a while but the boy had a lot of trouble with English. His wife came in near the end of the interview and cussed him out for answering the questions. She said, 'Don't you give him any information' about four times. After some harsh Spanish I didn't understand, she left and all the kids laughed. I was excitedly digging for my identification but the kids laughed and said not to mind her and to go on. R. left the room several times and I had to go and call him back, each time telling him that it would only take a couple minutes more. R. would pause for minutes at a time, then say, 'I don't know.'" (Respondent was Mexican-American)

"Very smoothly. There were no interruptions. The respondent understood all questions well. Very cooperative and eager to help out. The respondent sat in the chair in a very relaxed position. Her hair was pincurled. Seemed anxious that she answer the questions per se." (Respondent was Anglo)

The Coding Process

During the course of interviewing in 1959 preliminary codes were being developed and revised by pilot-coding a sample of completed interviews. Separate but sufficiently-similar-to-permit-comparison code books were constructed for the Mexican-American and Anglo samples. Upon completion of the interviewing a staff of 12 coders was trained. All schedules were coded, check-coded, and weak spots in the code detected and corrected. A study of coding reliability made it possible to promote the most efficient producers and retire those persons whose efforts were least reliable and least productive. These procedures yielded an average code reliability in excess of 90 percent. As an outgrowth of the coder reliability study and preliminary processing of the data, a number of the codes were collapsed in order to increase the reliability of predictorship categories and reduce coder variability to less than 10 percent.[3]

During the last week of interviewing in 1960, there were frequent periods when the interviewers had neither appointments nor persons on whom they might call to arrange appointments. Rather than have the interviewers idle during this period we decided to pilot-code a sample of schedules from each of the sub-populations in order to begin final coding earlier than otherwise possible. Pilot-coding in the field not only facilitated construction of the code book immediately upon completion of the interviewing but also reduced the amount of formal training required prior to beginning the final coding process. As in 1959, the staff included 12 coders. A 20 percent sample of the schedules was check-coded to determine coder reliability. Coding conventions developed during 1959 combined with the experience of that year enabled the 1960 interviews to be coded within five months. [4]

In 1961, work on the code book again commenced before interviewing was completed. By the time a majority of the interviews were completed the code book was ready and full-scale coding was in process. As in previous years, a 20 percent sample of the schedule was check-coded by the group and supervisors in order to insure coder reliability and a second sample was coded in order to measure column reliability, followed by collapsing of codes to achieve the level of column reliability desired.

Conclusion

Each year's survey gave us additional experience in the various tasks that were required in order to successfully interview the sample, code the data, and place it on IBM cards for preliminary analysis. Fourteen months elapsed between the first day of interviewing in Racine and the completion of machine runs with marginal totals in 1959, but only ten months were required for the same operation in 1960 and two months in 1961. The latter time was one-seventh of what it took in 1959 and one-fifth of what it took in 1960. A total of 1,524 interviews was processed. The results of our analyses will be presented in the following chapters.

NOTES

1 The use of tests of significance with survey research data has been questioned in recent years but has been vigorously supported as well. Although well aware of the controversy and some of the problems that are involved in defending the representativeness and independence of various samples that were selected from year to year, it was decided that the chi square test should be applied in testing the significance of differences between samples. Moreover, the chi square test has also been utilized in testing for differences between corresponding segments of two samples when it could be argued that each

segment is probably representative of that segment of the total population from which the sample was drawn.

While the decision was made to apply tests of statistical significance to the data it must also be emphasized that differences may be significant but at the same time be very small; there may either not be difference between two samples or the difference may be based on irregularities in the two distributions rather than on over-all directional differences. This is particularly true if the sample is large. On the other hand, rather large differences between samples may not be statistically significant if the sample is small. Statistical significance means only that the differences found between samples could only have occurred by chance one in so many times, depending on the confidence level selected or reported. For example, if a difference is significant at the .01 level it would have occurred by chance only once in 100 times. To summarize, significance tells us nothing about the strength of a relationship.

At this point, and in the next few chapters, we shall refer to the significance of differences. Then in later chapters the strength of our findings will be presented and discussed at length.

Although the literature dealing with the controversy is large, commencing with Hanan Selvin's "A Critique of Tests of Significance in Survey Research," American Sociological Review, Vol. 22, October, 1957, pp. 519-527, only a few additional articles need be cited for the reader who wishes to familiarize himself with the issues: Robert McGinnis, "Randomization and Inference in Sociological Research," American Sociological Review, Vol. 23, August, 1958, pp. 408-414; Thomas J. Duggan and Charles W. Dean, "Common Misinterpretations of Significant Levels in Sociological Journals," The American Sociologist, Vol. 3, No. 1, February, 1968, pp. 45-46; Denton E. Morrison and Ramon E. Henkel, "Significance Tests Reconsidered," American Sociologist, Vol. 4, No. 2, May, 1969, pp. 131-140; and Sanford Labovitz, "The Nonutility of Significance Tests: The Significance of Tests of Significance Reconsidered," The Pacific Sociological Review, Vol. 13, No. 3, Summer, 1970, pp. 141-148.

2. The problem of maximizing the commitment of all members of a research group to a project has plagued scientists in every field. Suggestions on how this may be facilitated are made in: Julius Roth, "Hired Hand Research," American Sociologist, Vol. 1, No. 4, August, 1966, pp. 190-196.

3. Acceptable coder variability meant that the coder and the check-coder had discrepancies not exceeding 10 percent on any single column; in reality the error was only 5 percent, assuming the coder to be correct half the time, and the check-coder the other half. Very few errors of a clerical nature were discovered and most coder variability was based on differences in judgment. In only a very few cases was collapsing unsuccessful in bringing coder variability to an acceptable level, and in cases where variability exceeded 10 percent, it was indicated in the completed code book containing marginal totals from computer runs.

4. A 5 percent sample of all schedules was check-coded for column reliability and average column reliability was 90 percent or better. As in previous years, the data on column reliability indicated that certain codes should be collapsed, thus making it possible to reduce coder variability to 10 percent or less on most columns.

Chapter 4

THE RELATIONSHIP OF SOCIAL ANTECEDENTS

TO ECONOMIC ABSORPTION

The Attitude Fallacy

Whenever a highly visible inmigrant group has not been absorbed into the economy to the same extent as have members of the larger or host society an explanation postulating ethnic or racial determinants may appear appropriate and sufficient to some persons in the larger community. Although interpretations of this nature in their simplest biological forms no longer have general credibility, explanations based on real or fancied "value differences" or differences in "world view" retaining racial and ethnic overtones may be considered adequate by persons with superficial knowledge of human behavior.

To have a "world view" which dictates short-term sacrifices in order to insure long-run or future benefits is so commonplace among middle socioeconomic status persons in the larger society that world views to the contrary are quite incomprehensible to them. They explain views that permit immediate gratification to the exclusion of long-range planning: (1) in terms of simple biological differences, i.e., that the inmigrant group is innately or by "nature" irresponsible or happy-go-lucky, or (2) by attitudes which have been learned or acquired so early in life that they have been translated into irreversible behavior patterns which make upward mobility virtually impossible.[1] More recently the latter interpretation has become part of the "culture of poverty" explanation. To be explicit, some who have not been able to view the problems of the less fortunate as innate or inborn have accepted the view that socialization in the subculture of poverty results in the acquisition of attitudes and behaviors that are almost as permanent and transmissible from generation to generation as if they were "inborn."[2]

What is more likely correct is that there is relatively little relationship between attitudes, values, world views, or behavior indicative of them and the extent to which the inmigrant has been absorbed into the economy. The Mexican-American's seeming disinclination to save does not explain the extent to which he has been absorbed into the economy; the type of job he is able to get simply makes it almost impossible to save for future contingencies.

At the same time that the urban middle-class observer employs explanations that are either simplistic on one hand or convoluted on the other, the inmigrant Mexican-American or Negro may be making an incorrect judgment about himself. His racial and ethnic explanation differs from that of the Anglo in that he perceives himself as someone who has been placed in a disadvantageous position by the larger community simply because he is a Mexican-American or Negro.[3] This perception of himself as basically disadvantaged because of race or ethnicity is correct if the "system" operates in such a way that employment opportunities for him are usually at the lowest levels or if it is difficult for him to associate with Anglos. The *degree* to which he is correct depends on the extent to which the larger community is organized to exclude the Mexican-American or Negro from participation in its activities.

A Better Point of Departure

Within what framework may we best explain that which impedes the Mexican-American's or Negro's absorption into the economy and integration into the larger culture of the northern industrial community? We hypothesize that an explanation that takes into consideration the interplay of the inmigrant's social antecedents and the nature of the society in which he is at the moment is more supportable than any biological, attitudinal, or currently popular systemic explanation of the discrimination type.

In this and the next chapter we shall concentrate on the presentation of a detailed analysis of the relationship of education and other sociological variables to the occupation and income level of Mexican-Americans, Negroes, and Anglos in Racine. Of Racine's total population of 89,144 in 1960, 28,038 were labeled "foreign

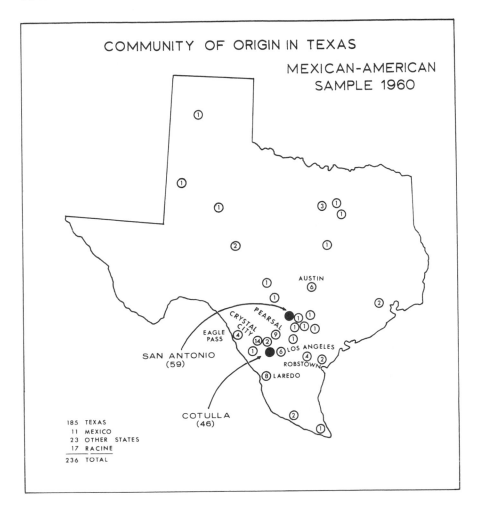

MAP 1

stock," of which 7,264 were classified as foreign born. The nonwhite (essentially Negro) population of 4,778, comprising only 5.3 percent was greatly underreported in the 1960 Census. There is, of course, no way to state with accuracy the number of Mexican-Americans in Racine at that time, but if we consider the school census of 1960 as a basis for estimation, the total would approach 4,000. The number of employed males in Racine was 23,170, of which 1,074 were given as nonwhite, also an underreported figure. In addition we can safely assume that 500 to 800 of the employed males were Mexican-American.

Place of Origin

Most of the Mexican-Americans in the 1960 sample were from the Southwest, while the Negroes were predominantly from the South (Maps 1 and 2). Of the 236 Mexican-Americans interviewed, 185 had come from Texas. Fifty-nine were from San Antonio and 46 from Cotulla; others, although scattered throughout the state, were mainly from the area south of San Antonio. Of the 280 Negroes in the sample, 128 were from Mississippi, with 39 from New Albany in the northeastern section and another group of 30 from Meridian and nearby small towns in the east-central portion of the state.

While over 50 percent of the Anglos had been born and always lived in Racine, 96 percent of the Mexican-Americans and 96 percent of the Negroes had moved to that city. The length of time that the Negroes had been in Racine was not significantly different from that of the Mexican-Americans (Table 4).

Although only 3 percent of the Anglo respondents were born outside the United States, about half their parents were born in Europe. Mexican-American respondents were most likely to have come from Texas or Mexico; in either case, Mexico was the most frequent place of origin for their parents. Of the Mexican-Americans interviewed in the 1959 sample, 23 percent engaged in migrant work or travel outside of Texas or Mexico on the way to the Racine area but only 26 percent had actually resided in a non-Mexican or non-Texas area before coming to Racine. When questioned about their former places of residence, some of the Mexican-Americans did not hesitate to discuss their illegal entry into the United States. "My mother carried me across the Rio Grande in her arms," or "We came in a rowboat," were typical responses of those who chose to go beyond the question put to them by the interviewer.

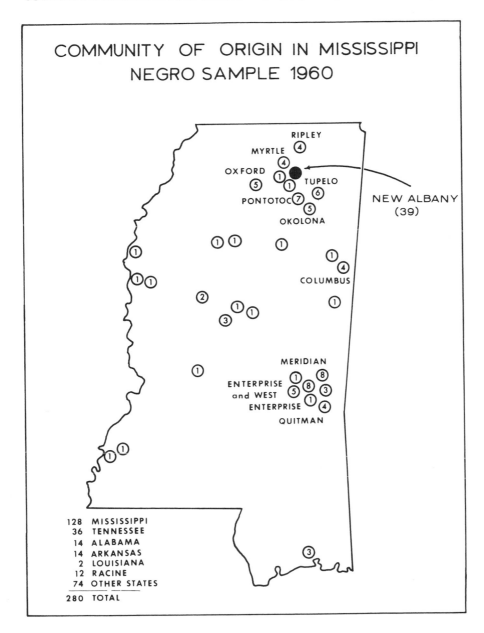

COMMUNITY OF ORIGIN IN MISSISSIPPI
NEGRO SAMPLE 1960

RIPLEY

MYRTLE ④

④

OXFORD ④

⑤ ①

TUPELO

PONTOTOC⑦ ⑥

⑤

OKOLONA

NEW ALBANY
(39)

① ①   ①

①

①

④

COLUMBUS

① ①

②

① ①

③ ① ①

①

MERIDIAN

ENTERPRISE
and WEST
ENTERPRISE

① ⑧

⑤ ⑧ ③

①

④

QUITMAN

① ①

128 MISSISSIPPI
 36 TENNESSEE
 14 ALABAMA
 14 ARKANSAS
  2 LOUISIANA
 12 RACINE
 74 OTHER STATES
280 TOTAL

③

MAP 2

TABLE 4. LENGTH OF RESIDENCE IN THE COMMUNITY

	Percentages		
Level of Education	Mexican-American	Negro	Anglo
0 years	17	1	1
1—2 years	9	0	0
3—4 years	21	6	2
5—7 years	20	21	4
8 years	9	18	15
9—12 years	13	38	60
13 or more years	1	3	17
Not reported	10	13	1
	100	100	100

The "not reported" category was eliminated for the purpose of computing chi square. Other categories were collapsed as necessary. Mexican-American/Anglo χ^2 = 252.38, 6 d.f., p < .001; Negro/Anglo χ^2 = 107.77, 4 d.f., p < .001; Mexican-American/Negro χ^2 = 123.40, 6 d.f., p < .001.

While the Mexican-Americans arrived in Racine by truck or automobile, the usual mode of transportation for them in the migrant cycle, the Negroes came by train or bus after stopping at various way-stations, such as Chicago and Kenosha.

Photographs of their communities of origin taken while traveling through the Southwest and the South fail to picture fully the way of life to which the inmigrants were accustomed before coming to Racine. Though we were fully cognizant of the "pull" of the northern industrial communities, the condition of the original homes of inmigrants vividly dramatized for us the "push" to leave that environment.

Rural or Urban Background

Though the general regional location of a respondent's place of origin is certainly an important antecedent, data revealing whether socialization took place within a rural or an urban context are likewise pertinent. Sociologists have traditionally taken the position that the rural-reared migrant commencing life in the hustle and competition of an industrial setting does so with a handicap that necessitates a period of transition and perhaps resocialization. A number of studies have shown that over a period of time some rural-reared migrants reach the levels of social participation of their urban counterparts, but that others do not.[4] But rural upbringing, though it may well be a handicap, is not of itself entirely insurmountable as evidenced by the successful absorption and integration into urban areas in the United States of millions of persons with rural backgrounds. The present study, as others, indicates that the economic absorption of persons with rural antecedents proceeds with time in the community, but that the effect of time in the community on the economic absorption of Mexican-Americans and Negroes, predominantly rural in origin, is limited by whatever occupational ceilings have been formally or informally established in the urban setting. In other words, the social organization of the urban community limits the influence of both time in the community and the level of aspiration of the Mexican-American or Negro inmigrant.

In measuring the rural-urban background of the three samples, population size of community of orientation was intended to be one index of the initial handicap of inmigrant workers. Size, although frequently utilized, is only one indicator of the kind of experiences that a person has had in a community during the process of socialization. For our purposes, the way of life for a group in that part of the country in which the community of orientation is located may be of even more significance. The question asked referred to "hometown" since "community of orientation" would not have been meaningful to most respondents. Unfortunately the data did not clearly reveal the size of community of orientation; when asked what they called "home" some of the respondents referred to nearby cities, either because it facilitated communication with the interviewer or because the name of a large city appeared more prestigious.

In some respects the best and most meaningful indicator of rural-agricultural versus urban-industrial background was occupation of the male's father. Not only did the fathers of the Mexican-American and Negro males have significantly lower-level occupations than did the Anglos, but the data on occupation of fathers (Table 5) clearly show that the modal background was agricultural for the Mexican-Americans and Negroes and industrial for the Anglos. Since no single measure of urban exposure was deemed adequate, the following Guttman scales measuring urban exposure or orientation were developed: (1) an index of the respondent's urban exposure prior to residence in Racine, and (2) an index of total urban exposure including length of time respondent had lived in Racine.[5] On both urban exposure scales, as shown in Table 6, Anglos had significantly higher scores than either Mexican-Americans or Negroes, while the differences between Negroes and Mexican-Americans were not significant.

TABLE 5. OCCUPATIONAL LEVEL OF FATHERS OF MALES

	1960 Percentages		
	Mexican-American	*Negro*	*Anglo*
Professional, technical, managerial, proprietor, clerical and sales	5	4	20
Craftsman, foreman	10	7	31
Operatives, maintenance and service, private household labor, industrial labor	20	24	28
Farmer and agricultural laborer	45	43	14
Not reported	20	22	7
	100	100	100

Mexican-American/Anglo χ^2 = 47.14, 1 d.f., p $<$.001; Negro/Anglo χ^2 = 85.11, 1 d.f., p $<$ .001; Mexican-American/Negro χ^2 = 1.52, 1 d.f., not significant; not reported excluded in calculating χ^2.

TABLE 6. PRIOR AND TOTAL URBAN EXPOSURE OF MALE RESPONDENTS

Scale Types		*Prior Exposure* *1960 Percentages*				*Total Exposure* *1960 Percentages*		
		Mexican-American	*Negro*	*Anglo*		*Mexican-American*	*Negro*	*Anglo*
Most urban	7	15	18	15	9	5	2	51
	6	24	8	59	8	29	24	21
	5	3	6	5	7	3	6	5
	4	8	9	1	6	11	9	2
	3	0	0	0	5	3	4	3
	2	26	43	10	4	21	25	11
	1	24	11	4	3	8	17	0
	0	0	5	6	2	21	5	1
					1	0	3	5
Least urban					0	0	6	1
		100	100	100		101	101	100

Prior Urban Exposure: [R = .9018; MR = .6328]; Mexican-American/Anglo χ^2 = 13.15, 1 d.f., p $<$.001; Negro/Anglo χ^2 = 49.17, 1 d.f., p $<$.001; Mexican-American/Negro χ^2 = 2.09, 1 d.f., not significant. Total Urban Exposure: [R = .9113; MR = .6789]; Mexican-American/Anglo χ^2 = 15.39, 1 d.f., p $<$.001; Negro/Anglo χ^2 = 45.26, 1 d.f., p $<$ .001; Mexican-American/Negro χ^2 = .70, 1 d.f., not significant. Sufficient data were available to develop scale scores for only 253 respondents. Differences on individual questions were of such an order as to indicate that these respondents are representative of the total samples.

The correlation of a single measure of urban exposure, such as size of community of orientation (birthplace or place of residence prior to coming to Racine), with present occupational level was low but statistically significant for Mexican-Americans but not for Negroes and Anglos. For the Negroes there was a low but statistically significant relationship between size of community of orientation and income. Age of male, length of respondent's time in Racine, and race and ethnicity were controlled in further analyses as an even more precise test of the hypothesis that the size of community of orientation is an important determinant of level of adjustment for the inmigrant in the urban community (as measured by occupational status). In none of the 12 tests of the null hypothesis was there any significant difference in occupational level of males that could be related to size, a fact which seemed to indicate that excessive emphasis has been placed on "sheer size" as a predictor of the level at which the inmigrant will either enter or ultimately attain in the urban industrial situation.

Actually, movement from rural to urban employment in itself assures some occupational mobility. An inmigrant formerly employed at the lowest levels in agriculture with a minimum amount of job security can usually obtain a job (if one is available) as an industrial laborer, in essence entering at the lowest level of the urban occupational hierarchy. Upon employment he is integrated into the urban industrial social welfare and security system; though he may never rise far within that occupational hierarchy he has moved into a system that provides relatively stable employment in contrast to that which he may previously have experienced. Since the level of entry and range of movement upward are, for a variety of reasons, limited for the Mexican-American and Negro inmigrant, the amount or length of his prior urban residence, no matter how it is measured, counts little in his struggle upward. In other words, prior urban residence per se does not help the inmigrant very much in the competitive world of urban industrial employment.

Occupational Background

It would seem that it is not so much prior urban experience as the amount of previous urban industrial work experience and other simultaneous work-related learning which might be related to the more successful absorption of inmigrants into the economy. The handicap of rural antecedents is minimal for some types of urban work. The former rural dweller who, because of his relevant experience, seeks work as a caterpiller tractor operator with a construction company will meet with more success than the one who attempts to secure a job in a furniture factory where quite different skills are required. By contrast, the inmigrant who has resided in an urban area in the Southwest but whose only experience has been as a migrant laborer has no preparation whatsoever for the work in the northern industrial community to which he is going. The complexity of the problem is such that some of the simpler rural-urban or city size relationships found in previous studies are of little help in understanding the processes of absorption into an urban industrial economy.

The relevant questions are, "What have been the first jobs of the respondents?" and "How are first jobs related to present jobs?" We are concerned not only with the occupational level of the respondent's first job but also with its regional location and the degree of urbanization of the job setting.

The first job held by an inmigrant seems to be closely tied to his subsequent level of employment in the industrial milieu. The social organization of the industrial community and its related subculture as well as the subculture of the surrounding region appears to influence persons in decision-making positions in the selection of particular kinds of persons for certain jobs. In time those who are selected come to expect this kind of work as the only work available. Though the assumption is prevalent that Mexican-Americans or Negroes like the kind of work they get, a more truthful description of the situation is that Mexican-Americans and Negroes have concluded (at least in the past) that since only certain types of labor are available to them, they will seek these jobs even though they may prefer a different kind, or higher pay, or those with regular employment. Thus it is that members of particular racial and ethnic groups tend to be channeled into lower-level positions in a community, especially if industrial work opportunities are relatively scarce, by what may be called the social system, the way the community is organized, or more simply, discrimination in employment practices.

First Job

The data in the samples showed that 52 percent of the Mexican-Americans and 47 percent of the Negroes, in contrast to 1 percent of the Anglos, began their employment experience in the

South or Southwest. When the size of the town where first job was held and the total years of urban employment exposure including employment in Racine were measured, the more urban nature of the Anglo work experience contrasted sharply with that of the Mexican-American or Negro.

The occupational level of first jobs of males is shown in Table 7. The category of farmer and farm manager is sometimes separated and both placed higher on the occupational level hierarchy but because of the size and nature of the farms from which the inmigrants came we established one category which included all persons engaged in agricultural endeavors. First jobs of Anglos were at significantly higher levels than those of the Mexican-Americans and Negroes but Mexican-American and Negro first jobs were not significantly different. It is very apparent that Mexican-Americans and Negroes did not have as much appropriate work experience as did the Anglos with whom they competed in the urban industrial area.

Employment in Racine

Mexican-Americans tended to be concentrated in foundries and other heavy industry. At the time of the 1959 survey 5 percent of the Anglos, as compared to 29 percent of the Mexican-Americans, were employed in foundries. In 1960 the Mexican-American unemployment rate, including unemployment due to strikes, was higher that that of the Negroes, and that of the Negroes higher than the Anglos. The susceptibility of Mexican-Americans and Negroes to

TABLE 7. OCCUPATIONAL LEVEL OF MALE'S FIRST JOB

	1960 Percentages		
	Mexican-American	Negro	Anglo
Professional, technical, managerial, proprietor, clerical and sales	3	4	25
Craftsman, foreman	4	7	17
Operatives, maintenance and service, private household labor, industrial labor	31	43	41
Farmer and agricultural laborer	37	28	8
Not reported	25	17	9
	100	99	100

Mexican-American/Anglo χ^2 = 91.71, 1 d.f., p $<$.001; Negro/Anglo χ^2 = 47.17, 1 d.f., p $<$ .001; Mexican-American/Negro χ^2 = 9.35, 1 d.f., p $<$.01; not reported excluded in calculating χ^2.

unemployment as a consequence of their position in the economic order is demonstrated by the fact that of those interviewed in 1960, only 61 percent of the Mexican-Americans were working, 3 percent were unemployed, and 28 percent were on strike; 69 percent of the Negroes were working, 4 percent were unemployed, and 16 percent were on strike; but 93 percent of the Anglos were working, 1 percent was unemployed, and only 4 percent were on strike. The influence of strikes on Negro unemployment becomes more apparent when we note that the strike in Racine's principal heavy industry had been settled by 1961 and the percentage of unemployed and not working among Anglos and Negroes was the same, 3 percent of each.

When hourly wages were compared, Anglo wages were significantly higher than Negro wages and these in turn were significantly higher than Mexican-American wages. Although for the most part Anglos had one job with a considerably longer work week than most Negroes or Mexican-Americans, in many instances Negroes and Mexican-Americans had multiple jobs, each of which had a short work week. Subsequently we found that the average Negro or Mexican-American spent far longer hours working per week than did the Anglo. Although 80 percent of the Anglos were working on the same job in 1961 as in 1960, a comparable measure of job stability or mobility for the Negroes was difficult to obtain since 29 percent of the Negro women did not know whether their husbands held the same job in 1961 as in 1960.

As emphasized earlier, the ultimate purpose of this volume is not merely to describe the inmigrants antecedents and present levels (1960) of economic absorption of Mexican-Americans, Negroes, and Anglos. We are concerned with a much more complex problem: how does the society operate so as to increase the likelihood that Mexican-Americans or Negroes will initially be located at lower levels in the occupational hierarchy than Anglos and, in spite of migration designed to better their position, continue to remain at substantially lower levels than the Anglos though they may have individual initiative and some opportunity? Studies of a more or less static nature have some merit but do not tell us how the society really functions. Those that do have a more dynamic aspect tend to emphasize what happens to a member of the dominant white majority rather than members of minority ethnic or racial groups in the larger society.[7]

Relationship of First Job to Present Job

The extent to which occupational antecedents may be determinants of the level of economic absorption in Racine is revealed when the relative amount of mobility between first and present jobs of the three groups is compared. Anglo first and present jobs were at significantly higher levels than those of Mexican-Americans or Negroes. For example, 27 percent of the Anglos had first jobs at the professional, technical, managerial, and proprietorial level while only 5 percent of the Negroes and 3 percent of the Mexican-Americans were at this level. Although Mexican-American and Negro first jobs did not differ significantly, Negro present jobs were significantly higher than those of the Mexican-Americans. The difference was particularly noticeable in the category of craftsmen and foremen where 26 percent of the Negroes but only 15 percent of the Mexican-Americans were to be found. The present job level of both Anglos and Negroes was significantly higher than that of their first jobs, but there was no significant difference in level of first and present jobs for Mexican-Americans (Table 8). The Mexican-Americans started as laborers and remained laborers. As previously stated, it is only when transition from agricultural labor to industrial labor is considered an upward movement that we can assert that Mexican-Americans have experienced significant mobility between first and present jobs.

Education and First Positions in the World of Work

The complex interrelationship of education with occupation and total family income, which we shall discuss in the next chapter also, was one of the most significant and interesting findings in the study. These findings merit serious consideration during a period when renewed emphasis has been placed upon both the desirability and the constitutional necessity of providing equal educational opportunities for all racial and ethnic groups, a concern that has in recent years resulted in U.S. Supreme Court decisions forcing movements in this direction in the South and more recently in the North. Data from the 1960 U.S. Census on the relationship of education to occupation have been presented ad infinitum as evidence that education is not only a powerful determinant of occupational level and income, but is in fact the key to success. Table 9 indicates that persons at the highest occupational level have the most education and that the amount of education of those at lower levels systematically

TABLE 8. FIRST AND PRESENT JOBS OF MALES—1960[1]

Occupational Level	Agricultural laborer	Industrial laborer	Maintenance and service	Operatives	Craftsman, foreman	Clerical and sales	Professional, technical, other	Total Percentage First Job
	I	II	III	IV	V	VI	VII	
FIRST JOB: MEXICAN-AMERICAN			PRESENT JOB: MEXICAN-AMERICAN					
I	4	44	1	18	10	1		48
II, III, IV	0	33	2	23	10	1		43
V	1	1	0	2	5	1		6
VI, VII	0	3	0	2	0	0		3
Total percentage	3	50	2	28	15	2		100
FIRST JOB: NEGRO			PRESENT JOB: NEGRO					
I	1	18	5	35	18	0		35
II, III, IV	3	34	9	45	26	0		52
V	0	0	1	5	12	0		8
VI, VII	0	3	1	3	2	3		5
Total percentage	2	25	7	39	26	1		100
FIRST JOB: ANGLO			PRESENT JOB: ANGLO					
I	1	1	1	5	9	3	2	9
II, III, IV	1	13	6	32	32	10	21	45
V	0	1	3	4	30	6	5	19
VI, VII	1	0	0	1	9	14	45	27
Total percentage	1	6	4	16	31	13	29	100

Mexican-American First/Present Job χ^2 = 9.30, 4 d.f., not significant; Negro First/Present Job χ^2 = 16.47, 4 d.f., p < .01; Anglo First/Present Job χ^2 = 88.68, 6 d.f., p < .001.

1. First and present jobs include only those respondents for whom data had been obtained on both first and present jobs. Grouped figures in boxes are raw frequencies, the marginal totals of each segment of the table are in percentages.

TABLE 9. THE EDUCATION AND OCCUPATION OF THE U.S. WORKING POPULATION (IN PERCENTAGES)

Occupational Level	Years of Education			
	Less than high school graduation	High school graduation	Some college education	Total
Professional and technical workers	6	19	75	100
Proprietors and managers	38	33	29	100
Clerical and sales	25	53	22	100
Skilled workers	59	33	8	100
Semi-skilled workers	70	26	4	100
Service workers	69	25	6	100
Unskilled workers	80	17	3	100
Farmers and farm workers	76	19	5	100

Source: Manpower-Challenge of the 1960's, U.S. Department of Labor, 1960, p. 17.

decreases. Although these data are for the total U.S. working population, it must be noted that the unemployed have been omitted and differentials by race, ethnicity, or region are not given. Because education is generally correlated with occupation and income, people are very often led to assume that there is a direct causal relationship between these factors and that a better formal education alone is *the* answer for those who wish greater incomes and higher level occupations. The Racine study, as other similar studies, emphasizes that it is the nature of the organization of society which dictates whether or not formal education can in fact be an equalizer of economic opportunities. The educational disadvantage of the Mexican-American (51 percent of the Mexican-Americans had received four or less years of education as compared to 7½ percent of the Negroes and 2½ percent of the Anglos, as shown in Table 10), when coupled with his language handicap, assumes even greater importance in understanding the extent to which he has not been integrated into the larger culture and absorbed into the economy.

But level of education *per se* is not the entire story; quality of education must also be considered. While 56 percent of the Mexican-American males and 88 percent of the Negro males received their education in the southern United States (including Texas), 95 percent of the Anglos received their education in the north-central states. Thirty-five percent of the Mexican-Americans and 56 percent

TABLE 10. STATED EDUCATION OF MALE RESPONDENT OR OF FEMALE RE-
SPONDENT'S SPOUSE AT TIME OF FIRST INTERVIEW

Level of Education	Percentages		
	Mexican-American	Negro	Anglo
0 years	17	1	1
1—2 years	9	0	0
3—4 years	21	6	2
5—7 years	20	21	4
8 years	9	18	15
9—12 years	13	38	60
13 or more years	1	3	17
Not reported	10	13	1
	100	100	100

The "not reported" category was eliminated for the purpose of computing chi square. Other categories were collapsed as necessary. Mexican-American/Anglo χ^2 = 252.38, 6 d.f., p < .001; Negro/Anglo χ^2 = 107.77, 4 d.f., p < .001; Mexican-American/Negro χ^2 = 123.40, 6 d.f., p < .001.

of the Negroes were educated in hamlets whereas 64 percent of the Anglos received their education in areas of 50,000 to 100,000 population. Eight percent of the Mexican-Americans and 15 percent of the Negroes as contrasted to 78 percent of the Anglos went to high school in the north-central section of the country, including Racine. One could go through the entire series of tables on the educational characteristics of the Mexican-Americans, Negroes, and Anglos and find essentially the same story, a picture of educational deprivation for the Mexican-Americans far exceeding that for the Negroes, and both worse than for the Anglos.

Tables 11, 11A, 11B, and 11C compare the educational status of the Racine samples with their as yet non-migrant counterparts or other appropriate groups in their community, county or state of origin and with all other persons in the United States. The detailed comparisons contained in Tables 11A, 11B, and 11C are presented in simplified form in Table 11. Despite some of the problems involved it is believed that such comparisons do give us a better understanding of how the inmigrant group may later come to perceive itself as either disadvantaged or discriminated against, or both, and how the Anglo residents of the community may perceive the educational disadvantage of the inmigrants to be a more important explanatory variable than it is.

The educational disparities that appear in Table 11 are precisely the ones most pertinent to our study. The most striking differences

TABLE 11. MEDIAN SCHOOL YEARS COMPLETED BY MALES 25 YEARS AND OVER, BY RACE OR ETHNICITY AND LOCATION

| | United States | Wisconsin | | Texas | | Mississippi | |
		State	Racine [3]	State	LaSalle County [6]	State	Union County [7]
Total	10.3	9.8	9.9 [4]	10.1	4.4	8.6	8.5
Mexican-American	7.1 [1]	—	4.8	4.8	1.4*	—	—
Negro	7.7 [2]	8.8	9.4 [5]	—	—	5.1	6.3*
Anglo	10.7	9.9	11.0	11.3	10.8*	10.6	9.0*

Sources: 1960 U.S. Census of Population: PC(1) 1C, Table 76; PC(1) 26C, Tables 47, 83, 87; PC(1) 45C, Tables 47, 83; PC(1) 51C, Table 47; PC(2) 1B, Tables 7, 14; U.S. Bureau of the Census, Current Population Report, Series P-20, No. 168, "Negro Population: March 1966," Table 6.

   Figures followed by asterisk(*) include both males and females, 25 years and over.

   1. Based on the number (3,464,999) of white persons of Spanish surname in the five southwestern states of Arizona, California, Colorado, New Mexico, and Texas which represents 87.5% of the total number (3,960,928) residing in the United States. Computed from data in **The Mexican-American People,** Leo Grebler, Joan W. Moore, Ralph C. Guzman, Appendix A, pages 606-607.

   2. The samples interviewed in Racine were "Anglo," "Negro," or "Mexican-American." However, the 1960 U.S. Census (the most reliable source beyond our own data) employs the terms "white," "nonwhite," and in some cases, "white persons of Spanish surname." It is well to remember that the term "nonwhite" could include those other than "Negro." In this specific instance, 7.7 is the median for the Negro population of the United States, while 7.9 is the median for the **nonwhite** population.

   3. Medians are for the samples interviewed in Racine.

   4. Given as 10.1 for **urban** Racine (1960 U.S. Census, PC(1) 51C, Table 73).

   5. Given as 8.6 for urban Racine, male and female (1960 U.S. Census, PC(1) 51C, Table 77).

   6. Twenty percent of the Mexican-Americans interviewed in the Racine sample came from Cotulla, Texas, located in LaSalle County. The population of the county was 5,972 (1960 U.S. Census, PC(1) 45C, Table 82), of which 3,832 are white persons of Spanish surname (64%) (1960 U.S. Census, PC(2) 1B, Table 15). The median for all persons 25 years and over in Cotulla is 4.4 (1960 U.S. Census, PC(1) 45C, Table 81).

   7. Fourteen percent of the Negroes interviewed in the Racine sample came from New Albany, Mississippi, located in Union County. The population of the county is 18,904, of which 3,311 (18%) are nonwhite (1960 U.S. Census, PC(1) 26C, Tables 82, 87), or in this case **Negro,** since there were no other nonwhites (1960 U.S. Census, PC(1) 26B, Table 28). The median for all persons 25 years and over for New Albany is 9.7 (1960 U.S. Census, PC(1) 26C, Table 81).

involve comparison of the inmigrant Mexican-Americans and Negroes with the Anglos in Racine, in particular the median of 4.8 years of education for Mexican-Americans as compared with 9.4 for Negroes and 11.0 for Anglos. Race and ethnic differences are neither so marked for the United States nor so noticeable between Negroes and Anglos in Wisconsin or Racine.

   Some interethnic differences in years of formal education in places of origin of the inmigrants were even greater than in Racine. In LaSalle County, for example, Anglos had a median of 10.8 years of education and Mexican-Americans 1.4 years. Not quite so disparate are the medians which are shown for Union County, Mississippi, where Negroes had a median of 6.3 and Anglos 9.0. Mexican-Americans were educationally disadvantaged in the Texas communities from which they came but continued to be educationally disadvantaged in Racine. Though the differences in years of formal education between Mexican-Americans and Anglos was less in Racine than in their places of origin, the nature of the urban industrial society of which the inmigrants became a part in Racine was such that, however less the difference, it was probably no less handicapping to them than existed in their former places of residence. Somewhat more complex were the educational differences between Anglos and Negroes. The Mississippi county selected as an example of one from which outmigration was taking place had less difference between Negroes and Anglos than the state as a whole. And in Racine the difference between Negroes and Anglos was not nearly so marked as that between Mexican-Americans and Anglos. Considering the quality of education available to Negroes who were educated in Mississippi before migration to Racine, the ability disparity might be greater than the medians indicate. With Table 11 as an introduction, we shall now turn to the remaining tables in the series for a more detailed picture of the educational qualifications of each of the Racine samples.

Looking at Table 11A we notice that the 1960 Racine sample of Anglos appears better educated than indicated by Census figures for

TABLE 11A. YEARS OF SCHOOL COMPLETED BY WHITE MALES 25 YEARS OR OLDER IN RACINE SAMPLE COMPARED TO UNITED STATES AND WISCONSIN, 1960 (IN PERCENTAGES)

Years of School Completed	United States		Wisconsin		Racine	
	Urban	Rural Non-farm	Urban	Rural Non-farm	City [1]	Sample
None	.2	2	1	1	2	1
1-4 years	4	8	4	6	4	2
5-7 years	12	17	12	16	12	4
8 years	17	20	22	29	23	15
9-14 years	43	39	42	36	46	60
13 or more years	23	14	20	12	14	17
Not reported	0	0	0	0	0	1
	101	100	101	100	101	100

Sources: 1960 U.S. Census of Population: PC(1) 1C, Table 76, PC(1) 51C, Tables 47, 73, 77.

1. Includes males and females 25 years and older. In this and other tables, city refers to the incorporated urban place.

urban Racine. Since Mexican-Americans were included in the Census figures for the white population of Racine but not in our sample of Anglos, the latter would be expected to have more education. When the Racine sample is compared with the urban and rural non-farm categories for Wisconsin and for the United States, this difference becomes even more noticeable. One may conclude that the white male population of Racine is fairly comparable to other white male urban populations in Wisconsin in respect to education, but that the Racine sample of Anglos has a larger proportion of those with some high school education.

Table 11B presents comparable data on years of education for the Racine sample of Negroes and other Negro populations. The

TABLE 11B. YEARS OF SCHOOL COMPLETED BY NEGRO MALES 25 YEARS OR OLDER IN RACINE SAMPLE COMPARED TO NONWHITES IN UNITED STATES, MISSISSIPPI AND RACINE, 1960 (IN PERCENTAGES)

| Years of School Completed | United States | | Mississippi | | | Racine | |
	Urban	Rural Non-farm	Urban	Rural Non-farm	Union County [1]	City [1]	Sample
None	5	12	9	13	5	2	1
1-4 years	17	31	30	42	30	12	6
5-7 years	22	25	26	25	36	24	21
8 years	13	10	12	9	15	19	18
9-12 years	34	18	18	9	12	39	38
13 or more years	10	4	5	2	3	4	3
Not reported	0	0	0	0	0	0	13
	101	100	100	100	101	100	100

Sources: 1960 U.S. Census of Population: PC(1) 1C, Table 76; PC(1) 26C, Tables 47, 87; PC(1) 51C, Table 77.
   1. Includes males and females 25 years of age and older.

distribution of the Racine sample was similar to that of the 1960 Census data for Racine and U.S. urban Negroes, assuming that the 13 percent for whom no information was obtained would have fallen in the lower educational levels. However, respondents in the Racine Negro sample claimed better educational qualifications than those for urban United States, and far superior ones than reported for their counterparts in Union County, Mississippi, rural and non-farm and urban Mississippi, and in rural and non-farm United States.

The data on Mexican-Americans presented in Table 11C are not as adequate for comparative purposes as were those for Anglos and

TABLE 11C.   YEARS OF SCHOOL COMPLETED BY MEXICAN-AMERICAN MALES 25 YEARS OR OLDER IN RACINE SAMPLE COMPARED TO SPANISH SURNAME POPULATION OF FIVE SOUTHWESTERN STATES, TEXAS, AND LASALLE COUNTY, TEXAS, 1960 (IN PERCENTAGES)

Years of School Completed	5 SW States	Texas	LaSalle County [1]	Racine
None	15	22	47	17
1—4 years	21	29	30	30
5—7 years	19	21	14	20
8 years	11	7	2	9
9—12 years	25	16	5	13
13 or more years	8	5	2	1
Not reported	0	0	0	10
	99	100	100	100

Sources: "Subject Reports Persons of Spanish Surname," Census of Population 1960, PC(2) 1B, Tables 7, 14.
    1. Includes Spanish surname female population 25 years of age and over.

Negroes, but they do give us some additional perspective on the difficulties of Mexican-Americans in Racine. The Racine sample of Mexican-Americans did not have as many persons in the highest educational levels as did the Spanish surname population of the five southwestern states (Texas, New Mexico, Arizona, California, and Colorado) or the population of Texas. On the other hand, the Racine sample did not have as large a proportion of unschooled persons as the Mexican-American population of LaSalle County and had more at higher levels. Even if we include in the calculation at the lowest levels those for whom education was unreported, the education of Mexican-Americans who moved to Racine would be far superior to those who remained in LaSalle County and probably superior to that of Mexican-Americans from other counties in Texas with high rates of outmigration. One could conclude that Mexican-Americans in Racine are similar in educational level to the Spanish surname population of Texas but probably superior to most Mexican-Americans remaining in the Texas areas from which our Mexican-American migrants came.

Occupation, income, and education are correlated, as shown in numerous analyses based on Census data, but we are most concerned with the extent to which formal education is a determinant of occupational level and income in specific regional and industrial settings and from one segment of society to another. Moreover, some individuals have incomes and occupations that are attributed to education but which should more appropriately be credited to such commonplace variables as membership in the so-called middle class, the acquisition of initial economic advantage, or simply being reared in an industrial or commercial center where education and employment opportunities abound. The "influence" of education not only commences at the moment when the respondent enters the social order but also operates in concert with other antecedent variables. Perhaps the point can be illustrated more clearly. When the decision to hire or not to hire is made, it is incorrect to assume that all gate-keepers have the same criteria in mind as they react to applicants. Is the gate-keeper looking for someone with a particular kind of work experience or skill? The data from this and other studies suggest that when an employer considers Mexican-Americans and Negroes as potential workers he places them in a category where the question of education and skill is of little consequence. Because the amount of education or skill that the applicant has is of secondary importance in successfully securing or holding the job, there is little likelihood that an *educated* Mexican-American or Negro will be considered for the same range of jobs as is the Anglo. When the Anglo encounters the gate-keeper his entire repertory of skills is taken into account unless he has readily visible defects or handicaps, and he is placed in an appropriate position commensurate with his education and skill.

The consequences of this process appear in Table 12 where the relationship between educational level and first job is presented. Whether Mexican-American, Negro, or Anglo, the higher the educational level attained *up to* eight years of education the higher the occupational level at which males entered the work order. But, among those with eight years *or more* of education, Anglo first jobs were at significantly higher levels than Mexican-American or Negro first jobs. Higher education had an initial value that was greater for Anglos than for Mexican-Americans or Negroes.

Once a person has entered the economy, the question becomes one of the relationship of educational antecedents to upward mobility in the occupational hierarchy. Have lack of education and lack of appropriate work experiences served as impediments to employment at higher levels and have these impeding factors had a greater influence on the economic absorption of Mexican-Americans and Negroes than on Anglos? Or, have race and ethnicity overshadowed education? There is every indication that the latter is so.

TABLE 12. EDUCATION AND FIRST JOB OF MALES IN RACINE SAMPLES, 1960

PERCENTAGES

	No education to four years			5-7 years			8 years			9-12 years			13 or more years		
	M	N	A	M	N	A	M	N	A	M	N	A	M	N	A
Professional, technical, managerial, proprietor, clerical and sales	2.	0	0	4	0	0	5	4	5	3	7	22	2	25	60
Craftsman, foreman	2	0	14	7	3	8	0	10	14	10	9	22	0	0	8
Operatives, maintenance, and service, industrial laborer	22	43	29	50	45	75	33	41	55	39	53	42	0	63	21
Farmer and agricultural laborer	51	57	43	17	43	8	29	33	12	23	20	7	66	13	2
Not reported	23	0	14	22	8	8	33	12	14	26	11	8	33	0	8
	100	100	100	100	99	99	100	100	100	101	100	101	101	101	99

Mexican-American χ^2 = 21.39, 4 d.f., p < .001; Negro χ^2 = 19.73, 5 d.f., p < .01; Anglo χ^2 = 33.74, 6 d.f., p < .001. (The not reported category for education included 1% of the Anglos, 10% of the Mexican-Americans, and 13% of the Negroes.)

NOTES

1. Negroes, of course, have their own stereotypes of what white people are like, with lower class Negroes having more intensely unfavorable conceptions of whites than middle or upper class Negroes. For an example, see: Paul A. McDaniel and Nicholas Babchuk, "Negro Conception of White People in a NOrtheastern City," Phylon, Vol. XXI, No. 1 (Spring, 1960), pp. 7-19. For an example of Mexican-American stereotypes of Anglos and self-images that form a framework for their day-to-day perceptions, see: Anthony Gary Dworkin, "Stereotypes and Self-Images Held by Native Born and Foreign Born Mexican-Americans," Sociology and Social Research, Vol. 49, No. 2, January, 1965, pp. 214-224. Further references will be made to Negro and Mexican-American self-concepts in several of the chapters that follow.

2. This is not to deny that there are important subcultural or attitudinal differences between groups; the literature is replete with examples of Negro-white differences. Glenn and Broom analyzed ten national public opinion surveys and found differences that cannot be explained by educational level or regional distribution. But as they have pointed out, these differences may be explained by variation in the experiences and conditions of Negroes. Although the average Negro-white response differentiation was less than differences between whites by region and education, the authors concluded that there is a unique Negro subculture in the United States. Leonard Broom and Norval D. Glenn, "Negro-White Differences in Reported Attitudes and Behavior," Sociology and Social Research, Vol. 50, No. 2, January, 1966, pp. 186-200. The culture of poverty concept will be examined at some length in Chapter 9 on World View. Suffice it to say at this point that Oscar Lewis should probably be credited for first suggesting this concept, at least in recent times, in his case studies of Mexican families, Five Families, New York: Basic Books, 1959. Ulf Hamerz's recent volume on the black ghetto in Washington, D.C., Soulside: Inquiries into Ghetto Culture and the Community, New York: Columbia University Press, 1969, indicates an awareness of the implications of substitution of the "culture of poverty" explanation for biological explanations.

3. A chapter will be devoted to attitudes of Racine's Negroes and Mexican-Americans towards their position in the larger society but it should be noted at this point that what we shall later report is not inconsistent with more current studies of racial attitudes. See for example, Angus Campbell and Howard Schuman, Racial Attitudes in Fifteen American Cities, Survey Research Center, Institute for Social Research, The University of Michigan, Ann Arbor, 1968. When asked a series of questions on discrimination, 72 percent of the males stated that they believe many or some Negroes miss out on jobs today because of discrimination and 68 percent stated that they believe many or some Negroes miss out on promotions today because of race.

4. The post-World War II to 1965 literature on the adjustment of migrants has been reviewed by Lyle W. and Magdaline W. Shannon, Chapter 2, "The Assimilation of Migrants to Cities," Leo F. Schnore and Henry Fagin (eds.), Urban Research and Policy Planning, Vol. I, Urban Affairs Annual Review, Beverly Hills: Sage Publications, 1967 and Chapter 2 of a paperback revision of the same, Leo F. Schnore (ed.), Social Science and the City, New York: Frederick A. Praeger, 1968.

5. The male respondent who had been exposed the least to an urban environment prior to moving to Racine had the following characteristics while the person with the next highest score had each of these characteristics except the last, i.e., went to high school in a city with population of 10,000 or more, and so on.

(a) Had not held a job in an urban place prior to coming to Racine.

(b) His birthplace had a population of less than 10,000.

(c) The place where he went to elementary school had a population of less than 10,000.

(d) His hometown had a population of less than 10,000.

(e) His last place of residence had a population of less than 10,000.

(f) He had spent nine years or less in last place of residence.

(g) The place where he went to high school had a population of less than 10,000.

The male respondent who had the least total urban exposure including Racine had the following characteristics while a person with the next highest score had all except the last item, i.e., had spent six years or more in a job in an urban place including Racine.

(a) Had moved to Racine.

(b) His birthplace had a population of less than 10,000.

(c) The place where he went to elementary school had a population of less than 10,000.

(d) His hometown had a population of less than 10,000.

(e) His last place of residence had a population of less than 10,000.

(f) He had been in Racine nine years or less.

(g) He had spent nine years or less in last place of residence.

(h) The place where he went to high school had a population of less than 10,000.

(i) He had spent five years or less in a job in an urban place including Racine.

The coefficient of reproducibility R, and the minimum coefficient of reproducibility MR, will be presented for each scale. The coefficient of reproducibility should be at a minimum level of .900 and the minimum coefficient of reproducibility should be considerably less if the scale is to be much of an improvement over marginal reproducibility. To the extent that scale scores permit the reproduction of response or data patterns beyond what may be done from the marginal totals, then the scale permits the representation of patterns with greater accuracy than would be possible from the marginals.

6. The location of this all-inclusive agricultural category is similar to its position in that order of estimated desirability for various categories of persons in agriculture presented by Norval D. Glenn in "Some Changes in the Relative Status of American Nonwhites, 1940 to 1960," Phylon, Vol. XXIV, No. 2, Summer, 1963, p. 110. Glenn's ranking of 11 occupational categories according to estimated desirability was based on the income and education of experienced workers in the occupational group and differs in several other respects from the seven-step hierarchy utilized in most of the tables in this volume. Arguments can be made for either system but that which we have employed is sensitive to the upward mobility of migrants without consisting of so many categories as to be unwieldy.

7. Many descriptive articles have a dynamic aspect in that they suggest that chains of events, or interrelated variables, or systematically made decisions count up to more than single factors in preventing members of a group from moving upward in the social system as might be expected considering their qualifications. These articles usually cite examples of the phenomena to which they refer but do not have the systematically gathered data that permit one to describe the operation of the social system with the precision that is desired. For an article depicting the complex nature of the problem, see: Paul Bullock, "Employment Problems of the Mexican-American," Industrial Relations, Vol. 3, No. 3, May, 1964, pp. 37-50.

Chapter 5

THE RELATIONSHIP OF SOCIAL ANTECEDENTS TO

PRESENT OCCUPATION AND INCOME

Differences in Unemployment Rates

Any attempt to compare either the occupational level or income level of two or more groups in several locations must take into consideration variations in rates of participation in the labor force and rates of unemployment or partial employment. Of primary importance, therefore, are the regional and race or ethnic differences in unemployment shown in the tables that follow.

Table 13 provides a comparison of unemployment rates by areas and race or ethnic groups. Taking everything that we know about Census data into consideration, particularly what one must do to be classified as not "at work" but looking for work, it is probably correct to say that Negro and Mexican-American unemployment is at least twice that of Anglos, even though that degree of difference is not quite shown for the United States as a whole in Table 13.[1] Indeed, the organization of society makes it difficult for the southern Negro to qualify as unemployed unless he resides in an urban industrial area.

The fact that unemployment was at least three times as high in Wisconsin for Negroes as for Anglos (twice as high in Racine), where reporting is likely to be more accurate than in the larger United States, tends to support our estimate of race and ethnic differentials. Here we should recall that in the survey sample 3 percent of the Mexican-Americans were unemployed while 28 percent were on strike, 4 percent of the Negroes were unemployed with 16 percent

TABLE 13. PERCENTAGE OF UNEMPLOYED MALES, 14 YEARS AND OVER, CIVILIAN LABOR FORCE, BY RACE OR ETHNICITY AND LOCALE, 1960

| | United States | Wisconsin | | Texas | | Mississippi | |
		State	Racine	State	LaSalle County	State	Union County
Total	5.0	3.9	4.2	4.4	9.3	5.4	4.2[5]
Mexican-American	8.0[1]	—	—[2]	8.2[1]	13.0	—	—
Negro	8.8	11.4	7.9	7.3	—[3]	7.1	8.1
Anglo	4.6	3.7	4.0	4.0	4.3[4]	4.5	3.6

Sources: 1960 U.S. Census of Population, PC(1) 1C, Table 83; PC(1) 51C, Tables 53, 73, 77; PC(1) 45C, Tables 53, 81, 83; PC(1) 26C, Tables 53, 81, 83, 87; PC(2) 1B, Tables 6, 14.
   1. Based on white persons of Spanish surname, five southwestern states.
   2. Mexican-Americans and other white persons of Spanish surname included under whites.
   3. There were very few nonwhites in LaSalle County. 1960 U.S. Census, PC(1) 45B, Table 28.
   4. White persons minus those of Spanish surname.
   5. There were only whites or Negroes in Union County. 1960 U.S. Census, PC(1) 26B, Table 28.

on strike, but only 1 percent of the Anglos were unemployed and only 4 percent were on strike.[2] Although the organization of society has placed Mexican-Americans and Negroes in positions of job instability to a far greater extent than for Anglos, we must conclude that the U.S. Census data on unemployment scarcely revealed the total picture.[3]

It appears as if essentially the same racial and ethnic disparities are found in Texas as in Wisconsin. Furthermore, Texas had the highest percentage of unemployed Mexican-Americans in the Southwest in 1960. In LaSalle County, Texas, the gap between Anglo and Mexican-American unemployment was quite startling. In fact, the sheer percentage of unemployed Mexican-Americans in Cotulla[4] defined it and LaSalle County as a place from which Mexican-Americans would literally be pushed in a search for employment. Between 1950 and 1960 the county had a decline of 24 percent in Spanish surname persons as compared with an overall increase of 38 percent for Texas; of the 255 counties in the state, only ten others had a sharper reduction.

The Census data for Mississippi in 1960 probably did not indicate the actual state of affairs there any more than in Texas, perhaps less. While the Negro/Anglo difference was pronounced, the nature of employment is undoubtedly such that underemployment or partial employment would better characterize the actual situation of a large

TABLE 14. EMPLOYED MALES CLASSIFIED OPERATIVES OR ABOVE, BY RACE OR ETHNICITY AND LOCALE, 1960 (IN PERCENTAGES)

	United States		Wisconsin			Texas[5]		Mississippi[7]		
	Total	Urban	State	Urban	Racine[3]	State	Urban	State	Urban	Union County
Total	74	80	72	84	85	71	78	58	78	48
Mexican-American	55[2]	63[2]	–	–	41[4]	54[2]	61[2]	–	–	–
Negro[1]	45	50	60	60	69	38	41	30	51	19
Anglo	77	83	72	84	86	78[6]	86[6]	73	89	53

Sources: 1960 U.S. Census of Population: PC(1) 1C, Tables 87, 88; PC(1) 51C, Tables 57, 58, 74, 78; PC(1) 45C, Tables 57, 58; PC(1) 26C, Tables 57, 58, 84, 88; PC(2) 1B, Table 6; PC(2) 1C, Tables 32, 55.

1. All figures given are for Negroes, except for Racine and Union County, which are for nonwhite.
2. Percentages for five southwestern states (Texas included) based on total white persons of Spanish surname.
3. Racine refers to incorporated urban place.
4. Percentage is for the Mexican-American sample interviewed in Racine. The percentages of other samples interviewed in Racine are: Negro—57% and Anglo—86%.
5. The percentage of all employed males as operatives or above is 39% for LaSalle County and 52% for Cotulla [PC(1) 45C, Tables 84, 81].
6. These figures are for Anglos, rather than whites.
7. The percentage of all employed males as operatives or above is 63% for New Albany [PC(1) 26C, Table 81].

proportion of the population. The difference between Negroes and Anglos was even greater for Union County than for the state.[5]

Were we to add all of the unemployed to the lowest level of the occupational hierarchy, for that is almost surely where they would go, we would have a better picture of the comparative occupational status (level at which the male usually works) of Mexican-Americans and Negroes in the South and Southwest as contrasted to Wisconsin and particularly to Racine. Similarly, since median income data are based on persons with income, the differences between groups would become markedly greater if those without income were included in these computations.

Occupation

In Chapter 4 we compared the occupational ranking and the relationship of first job to present job of Racine's Mexican-Americans, Negroes, and the Anglo majority. In this chapter we are describing the process of change from one set of societal positions in the world of work (a major segment of people's lives) to another. Tables 14, 14A, 14B, and 14C present the occupational levels of the Racine samples compared with appropriate U.S. Census data for counterpart populations in the United States, Wisconsin, Texas, and Mississippi. Since it would be rather difficult to present a summary table of median occupational levels, Table 14 simply shows the percent of the labor force that is at the level of operative or above.

The occupational rank-ordering of Anglos, Mexican-Americans, and Negroes as shown in any of these areas should not be surprising; Anglos rank considerably higher than Mexican-Americans, and Mexican-Americans are ahead of Negroes with but one exception. The *only* locale in which the Mexican-Americans fell below the Negroes is in Racine.[6] How much this may be attributed to the historical antecedents of the Mexican-American compared to that of the Negro is difficult to ascertain; we must be content with what may be learned through research about the influence of the acquired abilities and experiences of the inmigrants (before and after arrival in Racine) and the operation of such temporal and systemic variables as we are able to observe and measure as they relate to the process of economic absorption.

While the data are not always available to present comparisons involving the two communities (or the counties in which they are included) from which a considerable proportion of the inmigrants have come, we have included it in every possible instance. Cotulla (LaSalle County) and New Albany (Union County) may not be equivalent in all respects to the other communities from which inmigrants have come but they are more or less representative of Texas and Mississippi communities and are probably sufficiently representative for comparative purposes. The reader may have wondered why the year 1960 rather than 1950 or 1940 was chosen as indicative of the antecedents of the inmigrants. Our answer is that we may have chosen any point in time since the processes that are being described in this volume are continuing. Had we, for instance, utilized 1950 Census data for Racine and other locales, the nature of the differences would have been the same although the contrasts would probably have been more startling. In a sense, the antecedents to which we refer are experiences in a world of work in which jobs are low-skilled and wages are low. It is from this world of work that the inmigrant has come to a society in which production is organized so that skills are more in demand and wages are higher.

Anglos

The distribution of employed white males according to major occupational categories is presented in Table 14A in order to facilitate a more detailed comparison of the U.S. Census data including white Spanish surname persons with the Racine sample of Anglos. Although both the sample and the city had the same percentage of persons at the level of operative or above, the Racine Anglo sample was skewed toward the higher end of the occupational level scale more than was the employed white male population (1960 Census) of Racine. The latter differs little from that of the employed white male urban population of either Wisconsin or the United States. The Anglo sample interviewed in 1960 in Racine was a bit more representative of higher occupational levels than Racine's employed white population, or employed white urban populations in general. However, the Racine sample, consisting of only ever-married Anglos, should be skewed upward in comparison with the U.S. Census which includes both Anglos and Mexican-Americans under the classification of "white" and does not include the requirement of ever-married. The Racine sample may therefore be considered more representative of similar Anglo urban industrial populations.

TABLE 14A. OCCUPATIONAL DISTRIBUTION OF EMPLOYED WHITE MALES IN UNITED STATES, WISCONSIN, AND RACINE, 1960* (IN PERCENTAGES)

Occupation	United States Total	United States Urban	Wisconsin State	Wisconsin Urban	Racine City	Racine Racine[1]
Professional, techni- cal, managerial, proprietor	22	26	18	22	20	28
Clerical and sales	15	17	12	15	14	12
Craftsmen, foremen	20	21	19	22	25	30
Operatives	20	19	23	25	27	16
Maintenance and service	5	6	5	6	6	4
Industrial laborers	6	5	6	5	5	7
Farmers and agricul- tural laborers	8	1	14	1	0	1
Not reported	4	5	3	4	3	2
	100	100	100	100	100	100

Sources: 1960 U.S. Census of Population, PC(1) 1C, Table 88; PC(1) 51C, Tables 58, 74, 78.
  * Includes white Spanish surname except for Racine sample.
  1. Based on usual occupation of Anglos interviewed in sample, whether employed or unemployed.

Negroes

Comparison of the Racine sample of Negroes (Table 14B) with data from the U.S. Census for the city of Racine revealed that the former had disproportionately more craftsmen and fewer operatives, differences which became more pronounced when, according to the U.S. Census, we learn that Racine Negroes were more integrated into the industrial sector of the economy than were those for the entire United States. The difference is readily explained by the ever-married requirement for inclusion in the Racine sample. The occupational distribution of the Racine sample is skewed upward even more noticeably than that for Union County and the state of Mississippi; a greater percentage of the Racine sample Negroes were either craftsmen, foremen, or operatives (55 percent) as compared with the total Mississippi distribution (25 percent), urban Mississippi (42 percent), or the Union County distribution (16 percent). Whether reference is made to the Census data or to the sample, the higher level distribution of all Racine Negroes becomes more meaningful when the reader notes that 44 percent of the Negroes in Mississippi were farmers or agricultural laborers. That all Racine Negroes have been relatively more successful cannot be questioned; yet their

TABLE 14B. OCCUPATIONAL DISTRIBUTION OF EMPLOYED NEGRO MALES FOR UNITED STATES, MISSISSIPPI, UNION COUNTY, AND RACINE, 1960* (IN PERCENTAGES)

Occupation	United States Total	Urban	Mississippi State	Urban	Union County*	Racine City*	Sample
Professional, technical, managerial, proprietor	5	6	3	6	2	1	1
Clerical and sales	6	7	2	3	1	5	1
Craftsmen, foremen	10	11	7	13	3	12	22
Operatives	24	26	18	29	13	51	33
Maintenance and service	15	18	7	15	9	6	7
Industrial laborers	20	21	17	26	27	20	22
Farmers and agricultural laborer	11	1	44	4	37	0	1
Not reported	8	10	2	4	8	5	13
	99	100	100	100	100	100	100

Sources: 1960 U.S. Census of Population, PC(1) 26C, Table 88; PC(1) 51C, Table 78; PC(2) 1C, Tables 32, 55.
   *The nonwhite population of Union County was all Negro and was almost entirely Negro for Racine.

success is not so much greater than that for other urban Negro populations that they cannot be considered representative of Negroes who are being absorbed into the urban industrial milieu.

Mexican-Americans

The occupational distribution of the Racine sample of Mexican-Americans (Table 14C) was skewed more toward the lower end of the occupational level scale than that of Mexican-Americans in Texas, to the extent that Spanish surnames were indicative of the distribution of Mexican-Americans in Texas. The data for 1960 for Cotulla were for all persons; if the pertinent information for this predominantly Mexican-American city (64 percent) had been available it would probably have shown more Mexican-Americans skewed toward the bottom of the occupational hierarchy than for the Racine sample. It would be safe to say that most of the 20 percent classified as "professional, technical, managerial, or proprietor" in the 1960 Census were Anglo, although there have been varying numbers of Mexican-American places of business in Cotulla—cantinas, restau-

TABLE 14C. OCCUPATIONAL DISTRIBUTION OF EMPLOYED WHITE MALES OF SPANISH SURNAME IN FIVE SOUTHWESTERN STATES, TEXAS, AND RACINE SAMPLE, 1960 (IN PERCENTAGES)

	5 SW States		Texas			Racine
Occupation	Total	Urban	State	Urban	Cotulla	
Professional, techni- cal, managerial, proprietor	8	10	8	9	20	1
Clerical and sales	8	10	8	10	10	0
Craftsmen, foremen	16	18	16	18	11	14
Operatives	23	25	22	24	11	26
Maintenance and service	7	8	7	9	6	1
Industrial laborers	14	16	16	18	10	46
Farmers and agri- cultural laborers	18	8	19	7	26	4
Not reported	5	5	4	5	6	7
	99	100	100	100	100	99

Sources: 1960 U.S. Census of Population: PC(1) 45C, Table 81; PC(2) 1B, Table 6.

rants, and the like. With even more confidence we can assert that the 26 percent in agricultural labor were almost entirely Mexican-American. Eighty-six percent of Racine's Mexican-Americans were in industrial employment in contrast to the 32 percent of Cotulla's population that could at the most be engaged in industrial labor. Actually, industry in terms of northern industrial criteria does not exist in Cotulla; it is the breadth of this classification which explains any inclusion in it. We can see more clearly why Mexican-Americans have left Cotulla, since it is in industrial employment that they have had their greatest opportunity to become integrated into the larger society and absorbed into an economy which provides security of a type not obtainable in their communities of origin.

Although neither the Anglo nor the Negro samples may be quite representative of all Anglos and Negroes in Racine in terms of occupation, nor the data complete for the Mexican-American sample, these minor sampling questions should not in any way limit the validity of conclusions that may be reached in our analyses of the relationship of education to occupation.

Education and Occupation

Data from the 1960 Census clearly demonstrate the inexact

TABLE 15. DISTRIBUTION OF U.S. EMPLOYED WHITE AND NONWHITE MALES BY OCCUPATION AND YEARS OF SCHOOL COMPLETED, MARCH 1959 (IN PERCENTAGES)

| | TOTAL | | YEARS OF SCHOOL COMPLETED | | | | | | | |
| | | | 8 years or less | | 1-3 years High School | | 4 years High School | | College (1 or more) | |
	Non-White	White	Non-White	White	Non-White	White	Non-White	White	Non-White	White
Professional, technical, managerial, proprietors	6	26	2	11	2	15	5	23	45	63
Clerical and sales	6	13	2	6	6	10	17	19	20	18
Craftsmen, foremen	9	21	7	22	14	27	11	24	10	8
Operatives	24	19	24	25	32	27	23	18	5	5
Maintenance and service	14	6	11	8	17	6	23	5	14	2
Industrial laborers	27	6	34	10	23	7	15	4	5	1
Farmers and agricultural laborers	14	9	20	18	6	8	6	7	1	3
	100	100	100	100	100	100	100	100	100	100

Source: The Negroes in the United States: The Economic and Social Situation, U.S. Department of Labor, Bulletin No. 1511, June, 1966, Table IV B-10, p. 204.

relationship of education to occupation for employed "whites" and "nonwhites." The percentages at various occupational levels (as shown in Table 15) reveal that nonwhites were skewed toward the lower end of the scale until they reached the college-educated category and here, although skewed toward the upper end of the scale, they did not reach equality with the whites. The dramatic differences in occupational distribution of whites and nonwhites may best be seen by examining the "professional, technical, managerial, and proprietor" category, noting that four to seven times as large a percentage of whites are found in each years-of-school-completed category as compared with nonwhites, until the level of college education is reached. Additional perspective may be gained by examining the categories of "maintenance and service" workers and "industrial laborers" where nonwhites were found in disproportional percentages at every educational level.

We are specifically concerned, however, with the extent to which the occupational distributions of Mexican-Americans, Negroes, and Anglos differ in the Racine sample, particularly within educational categories, and how these differences compare with differences for the United States.[7]

The educational versus occupational level data in Table 16 contain essentially the same occupational categories as does Table 15 except years of education has been divided into more categories at the lower levels to show that Mexican-Americans and Negroes in the Racine sample were skewed toward the lower end of the educational continuum. Because of the problems attendant to comparison of national population statistics with samples, particularly when cross-tabulations are involved, we do so with caution. Though occupational level did vary with number of years of education within each racial or ethnic group, the relationship of education to occupation was by far the greatest among the Anglos.

Over two-thirds of the Anglos with 13 or more years of education were in the "professional, technical, managerial, and proprietor" category. A college education seemed to have a higher pay-off occupationally for the Racine Anglo sample than for all Anglos in the United States. This finding would be expected since education is more valued in an urban industrial area where specialized training is needed in the competition for professional and technical positions.

By contrast, not only were Negroes with some college education almost without exception at lower occupational levels than comparable Anglos in Racine, they were not doing as well as were their

TABLE 16. EDUCATION AND PRESENT OCCUPATION OF RACINE MALES, 1960 (IN PERCENTAGES)

LEVEL OF EDUCATION

	No education—4 years			5-7 years			8 years			9-12 years			13 or more years		
	M	N	A	M	N	A	M	N	A	M	N	A	M	N	A
Professional, technical, managerial, proprietor	0	0	14	2	0	8	0	0	15	7	0	21	0	12	71
Clerical and sales	0	0	0	0	0	8	0	0	2	0	2	15	0	0	12
Craftsmen, foremen	14	14	43	16	27	33	10	22	32	14	30	34	33	0	17
Operatives	21	38	28	34	39	33	45	50	22	48	38	18	0	50	0
Maintenance and service	2	5	0	0	5	8	0	9	7	3	8	5	0	12	0
Industrial laborers	55	38	14	48	27	8	40	17	15	28	21	6	66	25	0
Farmers and agricultural laborers	8	5	0	0	2	0	5	2	5	0	0	1	0	0	0
	100	100	99	100	100	98	100	100	98	100	99	100	99	99	100

Although it is possible to select cutting points that reveal a significant relationship between education and occupational level for Mexican-Americans, this may be done only by lowering the educational cutting point and occupational cutting point below that utilized for the Anglo tests. Shifting cutting points could in no way produce a significant relationship for the Negroes.

Mexican-American $\chi^2 = 9.89$, 1 d.f., p $<.01$; Anglo $\chi^2 = 13.61$, 1 d.f., p $<.001$; Negro $\chi^2 = 1.24$, 1 d.f., not significant. Nine to 12 years of education: Mexican-American/Anglo $\chi^2 = 22.69$, 1 d.f., p $<.001$ Negro/Anglo $\chi^2 = 34.11$, 1 d.f., p $<.001$; Mexican-American/Negro $\chi^2 = .95$, 1 d.f., not significant. Eight years of education: Mexican-American/Anglo $\chi^2 = 7.54$, 1 d.f., p $<.01$; Negro/Anglo $\chi^2 = 6.33$, 1 d.f., p $<.02$; Mexican-American/Negro $\chi^2 = .62$, 1 d.f., not significant. Less than eight years of education: Mexican-American/Anglo $\chi^2 = 13.44$, 1 d.f., p $<.01$; Negro/Anglo $\chi^2 = 2.37$, 1 d.f., not significant; Mexican-American/Negro $\chi^2 = 9.11$, 1 d.f., p $<.01$. N = 200 Mexican-American, 230 Negro, and 276 Anglo. Not reported excluded from table.

counterparts in the United States. Although we believe that this is an accurate portrayal of the employment picture in Racine, particularly since our sample is biased in favor of overrepresentation of Negroes at higher occupational and educational levels, we must be a bit cautious because the small number of Negroes in the sample with a college education may not be representative of Racine's college educated Negroes. Neither of the two Mexican-Americans with some college education had even reached the clerical category.

Among those with nine to 12 years of education, Anglos had significantly higher present jobs than either the Mexican-Americans or Negroes. The difference is so startling that one need only glance at the table to verify it. High school educated Anglos in the Racine sample were at higher occupational levels than in the United States as a whole; but Negroes, rather than being at the level that one would expect, were slightly lower.

Among those with eight years of education, the difference between Anglos and Mexican-Americans or Negroes was still significant, though not as great. The difference between Mexican-Americans and Negroes was not significant although the level of Negro jobs tended to be higher than that of Mexican-Americans. The only point at which Mexican-Americans and Negroes showed a significant difference, and where the Anglo/Negro relationship showed no significant differences, was among those with *less* than eight years of education. If there was any single statistic that showed the relative occupational success of Negroes in comparison with Mexican-Americans in this particular industrial community it was the finding that there were proportionately fewer Negroes than Mexican-Americans at the lowest occupational level (industrial laborer), particularly among those in the lowest educational levels.

Since Negroes, in comparison with Anglos, were relatively new arrivals in the community and the community had not developed the entire gamut of specialized services that would be found in an older Negro community, there was proportionately less pay-off for Negroes educationally than for Anglos when compared with the larger United States. Negroes, on the other hand, came to Racine before Mexican-Americans and this historical (temporal) factor has resulted in the development of sufficient community and social organization to facilitate absorption into the economy to a greater extent for them than for Mexican-Americans who had no specialized or separate segment of the economy that might be called their own.

TABLE 17. MEDIAN INCOME FOR FAMILIES, BY RACE OR ETHNICITY AND LOCALE, 1960*

	United States		Wisconsin			Texas			Mississippi		
	Total	Urban	State	Urban	Racine[3]	State	Urban	LaSalle County[6]	State	Urban	Union County[7]
Total	$5,660	$6,166	$5,926	$6,535	$6,758	$4,884	$5,331	$2,296	$2,884	$4,173	$2,274
Mexican-American[1]	4,165	4,466	–	–	4,746[4]	2,914	3,188	1,585	–	–	–
Negro	3,047[2]	3,631[2]	4,653	4,876	5,491	2,591	2,915	–	1,444	2,100	1,448
Anglo	5,893	6,433	5,950	6,577	6,812	5,636[5]	6,134[5]	4,784[5]	4,209	5,493	2,530

Sources: 1960 U.S. Census of Population: PC(1) 1C, Table 95; PC(1) 26C, Tables 65, 86, 88; PC(1) 45C, Tables 65, 86; PC(1) 51C, Tables 65, 76, 78; PC(2) 1B, Tables 5, 14; PC(2) 1C, Table 14.

* 1960 U.S. Census of Population uses income for 1959.

1. Median income for white persons of Spanish surname in five southwestern states. The median for all families in five southwestern states is $5,992; Anglo, $6,448; Nonwhite, $3,644. [Table 65 in each of PC(1) 4C, 6C, 7C, 33C, 45C; PC(2) 1B, Table 5.] The medians for Texas are also based on white persons of Spanish surname.

2. Median income of Negroes in U.S.; other medians are for nonwhite, though Mississippi has practically no other nonwhites than Negro.

3. Racine refers to incorporated urban place.

4. Median income is for Mexican-American sample interviewed in Racine; median income for Negro sample is $5,362 for Anglo, $6,638. "If not ascertained" are put in the lowest categories, the results are as follows: Mexican American, $4,627; Negro, $4,915; Anglo, $6,448.

5. Median incomes of Anglos rather than whites. Since there are practically no nonwhites in LaSalle County, it is possible to compute the Anglo median.

6. The median income for all families in Cotulla is $2,000 [PC(1) 45C, Table 81].

7. The median income for all families in New Albany is $3,436 [PC(1) 26C, Table 81].

Income

While the relationship between education and occupation is one of the best indicators of the extent to which a group has been absorbed into the economy commensurate with its qualifications, the relationship of education to income is of equal importance in describing the extent to which a group has been absorbed. Table 17 is a summary table of median incomes for each racial and ethnic group for the United States and for the specific areas with which we have been concerned. Tables 17A, 17B, and 17C present a more detailed comparison of the Racine samples with their counterparts in other areas.

Respondents were not always able to give adequate answers to the series of questions about total family income. Fortunately, specific questions on hourly wages, hours worked per week, weeks worked per year, and additional income made it possible to compute each family's total annual income. Since the data are presented in minimum intervals of $2,000, it is believed that every family falls in essentially the same income category into which it would have fallen had responses to the U.S. Census been available for comparison. For this reason we believe that we are justified in comparing the Racine sample data on income with the U.S. Census.[8]

Table 17 has been constructed so that it is as comparable to Table 14 as possible in terms of the populations and sub-populations which the data represent. As in Table 14 on occupational level, the rank-ordering of Anglo, Mexican-American, and Negro, from highest to lowest income, prevailed in every locale except Racine, whether for the total population or the urban population. Further, the differences between locales were so marked that the likelihood of income improvement through migration from Texas or Mississippi to a northern industrial area was considerable; it could well have been the deciding factor in the decision-making process of potential migrants in the South or Southwest or of those in the migration cycle.

To be sure, the existence of significant income differences between Negroes in Mississippi and Racine or Mexican-Americans in Texas and Racine is only part of the story, but to the extent that differences are communicated to potential migrants these rather large income disparities would be a very attractive and powerful incentive to migration with the intent to establish permanent residence.

Table 17 shows that Mexican-Americans in Texas had a median

family income of $2,914, and $3,188 in urban areas, while their counterparts in Racine had a median family income of $4,746.[9] But more pertinent, in terms of the ties between communities, was the $1,585 median income of Mexican-Americans in LaSalle County as contrasted to $4,746 in Racine. Negroes in Mississippi had a median family income of $1,444 (Union County $1,488), $2,100 in urban areas, while Negroes in Racine had a median family income of $5,491. With these disparities in mind, the inmigrant does have a basis for believing that migration will bring about an improvement in economic status, perhaps not an immediate change from unskilled kinds of work but certainly a change in how well he will be paid for working. No matter where he originates, the Mexican-American or Negro from the South or Southwest will undoubtedly move to a higher wage area in the urban industrial North.

Anglo

The annual income of Anglo families in the Racine sample (shown in Table 17A) was less skewed towards either end of the continuum than income for the total white population for the city of Racine, while the latter was similar to that for either the United States or Wisconsin, except that it was slightly more skewed toward the high end of the scale. Variation of the Racine sample data on income from that of the Census data for Racine is not a matter for serious concern, as we have previously mentioned in reference to occupa-

TABLE 17A. INCOME OF WHITE FAMILIES IN UNITED STATES, WISCONSIN, AND RACINE, 1960* (IN PERCENTAGES)

Income	United States		Wisconsin		Racine	
	Total	Urban	Total	Urban	City	Sample
$15,000 and over	5	6	3	5	19	10
$10,000–14,999	11	13	8	13		
$7,000–9,999	21	24	18	26	28	29
$5,000–6,999	24	25	23	29	30	36
$3,000–4,999	20	18	19	16	13	13
$1,000–2,999	14	11	17	9	8	4
Under $1,000	5	3	11	2	2	
Not reported	0	0	0	0	0	8
	100	100	99	100	100	100

Sources: 1960 U.S. Census of Populaton: PC(1) 1C, Table 95; PC(1) 51C, Tables 65, 76, 78.
   *Includes white Spanish surname except for Racine sample.

tional distribution. In respect to income, many of those not reported for Anglos were due to special circumstances making it "impossible" for the respondent to answer. Some of these cases would go at the upper end of the continuum while others would go at the lower end, resulting in an income distribution that is fairly close to that for a Racine population similar to our sample of Anglos. Since the relationship of the sample income data to other variables such as education is our concern—the peakedness of the Anglo distribution serves only to limit the possibility of sufficient income variation to produce the correlations that might be expected with other variables.

Negro

Comparison of the Negro sample for Racine with other appropriate Census distributions is relatively simple if one accepts "nonwhite" as practically synonymous with Negro. Table 17B indicates that the Negro sample had neither as many low nor high income families as did Racine. Difficulties experienced by the interviewers in securing income data from female respondents accounted for most of the unreported income, suggesting that while a sizeable proportion of the "not reported" category of 15 percent should be placed in the lowest income categories, others should be assigned at all other levels. The sample would then become less skewed toward the low end of the continuum and become more representative of the Racine population of Negroes as determined by the U.S. Census. In comparison with urban Negroes in the United States, the Racine sample is skewed considerably upward and probably differs little from the Racine population. A fact to re-emphasize is that Negro income exceeded Mexican-American income in Racine.[10] Racine Negroes, either from the sample or from the Census data, were immensely better off than their counterparts in rural or urban Mississippi or Union County, Mississippi, and in all probability the United States, whether rural or urban.[11] Since we are not focusing our attention on differences between Negroes and whites in Mississippi, it is perhaps irrelevant to present any detailed comparison of them in that state, but the fact that in 1959 the median family income for Negroes was $1,444 and that for whites was $4,209 sheds some additional light on how a move to Racine would be defined as a step toward increased real as well as relative income.

In summary, the Racine study focused on a sample of Negroes

TABLE 17B. INCOME OF NEGRO FAMILIES IN THE UNITED STATES, MISSISSIPPI, AND RACINE, 1960* (IN PERCENTAGES)

	United States		Mississippi				Racine	
Income	Total	Urban	State	Urban	Rural Non-farm	Union County[1]	City	Sample
$15,000 and over	1	1	0	0	0	0 }	7	2
$10,000–14,999	3	3	0	1	0	0 }	19	13
$7,000–9,999	8	10	2	2	1	1	34	34
$5,000–6,999	14	17	3	5	2	4	25	24
$3,000–4,999	25	28	12	21	9	12	10	12
$1,000–2,999	34	30	46	51	45	49	10 }	12
Under $1,000	16	11	37	21	42	35	5 }	5 } 15
Not reported	0	0	0	0	0	0	0	0
	101	100	100	101	99	101	100	100

Sources: 1960 U.S. Census of Population: PC(1) 1C, Table 95; PC(1) 26C, Tables 65, 88; PC(1) 51C, Table 78; PC(2) 1C, Table 14.
* Percentages for the United States and the Racine sample are for Negro families. Those for Mississippi and City of Racine can be assumed so, as the numbers of nonwhites other than Negro are negligible.
1. Union County, in which New Albany is located, is one of 620 counties with the lowest median income in United States.

who, as a consequence of their absorption into the industrial system, had probably moved up the income ladder further than the occupational ladder.

Mexican-American

Although it was not possible to secure data permitting comparison of the Racine sample of Mexican-Americans with either the Racine population of Mexican-Americans or Spanish surname population, several other populations may be compared with the Racine sample (Table 17C) in order to present some idea of how persons in the sample did differ from those who remained in Texas. In the category of family income of less than $3,000 per year we could probably place not only the 9 percent of the Mexican-American sample in Racine who reported this income but also a proportion of the 7 percent who did not report an income figure. But since many of these, as in the case of the Negro sample, were females who did not know their husband's income, only a portion of the not reported should go there. The income picture for Mexican-Americans in Racine was not good in comparison to that of Negroes and Anglos. But it compared favorably with that of other Mexican-American populations; 60 percent of the Mexican-Americans in Cotulla, 67 percent of the non-farm Mexican-Americans in Texas, and 47 percent

TABLE 17C. INCOME OF FAMILIES WITH SPANISH SURNAMES IN FIVE SOUTHWESTERN STATES, TEXAS, AND RACINE SAMPLE, 1960 (IN PERCENTAGES)

| | 5 SW States | | Texas | | | Racine |
| | | | | | LaSalle | |
Income	Total	Urban	State	Urban	County [1]	
$15,000 and over	1	2	1	1 }	1	0
$10,000–14,999	5	6	2	2 }		0
$7,000–9,999	13	14	6	7	2	11
$5,000–6,999	20	22	13	14	2	29
$3,000–4,999	26	26	27	29	14	41
$1,000–2,999	26	23	38	35	48 }	12
Under $1,000	9	8	14	12	32 }	
Not reported	0	0	0	0	0	7
	100	101	101	100	99	100

Sources: 1960 U.S. Census of Population: PC(2) 1B, Tables 5, 14.
1. LaSalle County, in which Cotulla is located, one of those counties in Texas with the lowest median income ($2,500 and under), is also one of 620 counties with the lowest median family income in the United States.

of the urban Mexican-Americans in Texas reported less than a $3,000 family income in 1959.[12] Thus while more than four times as many urban Spanish surname (the closest that we can come to Mexican-American income data) families in Texas were below $3,000 in annual family income as urban Anglos in Wisconsin, the Racine sample of Mexican-Americans was neither as skewed towards the upper end of the scale as was the Racine sample of Anglos nor as skewed towards the lowest end of the scale as were the Texas Spanish surname persons.[13] The relationship of ethnicity to income is further demonstrated by comparison of the median 1960 income of Spanish surname persons in Texas for males aged 25 and over, which was $2,400 (average years of school 4.8) with that of Anglos which was $4,768 (average years of school 10.8).[14]

The Importance of Income

The family income data for Mexican-Americans in Texas, and especially in Cotulla, placed the Texas counterparts of the Racine inmigrants at low income levels; likewise, the Mississippi data placed Mississippi counterparts of the Racine Negroes at lower levels, particularly in rural non-farm areas and in Union County. Neither the Census data nor any of our data enable us to say that the Mexican-Americans and Negroes who were living in Racine in 1960 and reported on their 1959 incomes would have been in the lowest family income categories had they remained in Texas or Mississippi, but they do reveal that the inmigrants in our sample, or in the city of Racine for that matter, were far ahead of their Texas and Mississippi counterparts. The reader is, of course, aware that these comparisons, however much they show differences between Mexican-Americans, Negroes, and Anglos, are gross comparisons and do not really show the finer points of difference in levels of living that are possible with a given stated income. Within the past few years professional journals as well as the public press have contained articles describing differences in the cost of groceries and other necessities in slum neighborhoods compared with other neighborhoods. Others have dealt with the rental market and durables bought on credit. Such economic considerations make the Negro and the Mexican-American, to the extent that they are more segregated than other poor, special cases of poverty.[15]

Although detailed individual and family income information was collected each year of the study with the intent of making comparisons from group to group and from year to year, there were some yearly variations based on differences in the types of samples selected or some unusual event, such as the 1960 strike. For example, the distribution of hourly wages for 1958 as reported in the 1959 survey was almost identical for Anglos and Mexican-Americans because the Anglo sample was selected from lower socioeconomic status areas contiguous to the areas in which Mexican-Americans resided, whereas in 1960, it will be recalled, the Anglo sample was fairly representative of the larger Anglo population.

Generally speaking, the interrelationships which we are investigating are neither influenced by year-to-year fluctuations nor by the kinds of differences that were found between our samples and the population which they represented in Racine. We are more interested in the nature of the relationship between variables and how the society operates than in the exact relationship of Mexican-American, Negro, and Anglo hourly wage levels to each other. There were distinctive differences each year that the study was conducted, but of greater importance was the fact that Anglos were always better off in every respect than Negroes, and Negroes were better off than Mexican-Americans.

The Anglo work week was significantly different from that of Negroes or Mexican-Americans, but that of Negroes and Mexican-Americans was similar. One factor increasing the complexity of the data was that a large percentage of the Anglos, though regularly employed, were not held to fixed eight-hour days, five days per week. Instead, they worked somewhat irregular schedules six or seven days of the week, taking time off whenever they wished, or working from only a few hours to as many as 18 hours per day as production or business demanded it, without suffering any reduction in income. By contrast, when a Negro or Mexican-American had fewer work hours per week a reduction in pay was involved. He was also more likely than the Anglo to work longer than the normal work week of 40 hours by working at more than one job. The 1961 sample specifically shows that those Negroes from whom we obtained a work schedule tended to work longer hours than Anglos. But the occupational level at which the Negro labored was such that, even with the addition of longer hours at the same job, or at an additional job, his total family income was less than that of an Anglo.

To summarize all that has been said about income in each of the groups, Anglos had significantly higher incomes than the Negroes, and the income of the Negroes was significantly higher than the Mexican-Americans. There were differences in income within each level of occupational grouping with the Anglos higher than the Negroes and the Negroes higher than the Mexican-Americans. However, the difference in total annual income of each ethnic sample was not as great at the lower occupational levels as at the higher occupational levels. These differences were in part due to the breadth of each occupational category which included considerable variation in the exact type of work performed by each ethnic or racial group.

Education and Income

The relationship of education to income may be approached in several ways. One might be to explore the extent to which this relationship exists in Texas, Mississippi, and Wisconsin in comparison with Racine for each of the racial and ethnic groups. Unfortunately, as with occupation, it is difficult and sometimes impossible to make exact comparisons and we must settle for a general idea of the relationship of education to income as a basis for judging whether or not education has a greater pay-off for Mexican-Americans and Negroes in Racine than in the places from whence they came.[16]

In Table 18, median income is presented for several achieved educational categories. In the urban white population of Texas there was a definite upward progression of median income with education. When Spanish surname persons are independently observed the progression is not so marked, though median incomes were similar to the whites at lower educational levels. Obviously, increased increments of education generated proportionately less income for Spanish surname persons in Texas than for Anglos. However, these figures may be somewhat misleading for comparison with Racine since the Mexican-Americans who have moved to Racine comprise only a part of the Spanish surname population in Texas. That segment of the Spanish surname population in Texas which probably contributes heavily to the relationship between education and income for Spanish surname persons that does exist is the Hispano element which may or may not define itself as part of the Mexican-American population. Were it possible to compare those Mexican-Americans originally in Texas or Mexico a generation ago with Anglos, the differential pay-off of education for Mexican-Americans would be more apparent.

TABLE 18. EDUCATION AND MEDIAN INCOME OF URBAN MALES, 25 YEARS AND OVER IN TEXAS, MISSISSIPPI, AND WISCONSIN, 1960

MEDIAN INCOME

School Years Completed	Texas		Mississippi		Wisconsin	
	White [1]	Spanish Surname	White	Negro [2]	White [3]	Non-white [4]
No school	$1,525	$1,531	$1,039	$ 894	$1,839	$2,441
1-4 years	2,267	2,208	1,698	1,358	2,290	3,322
5-7 years	3,404	2,795	2,642	1,767	4,007	3,911
8 years	3,998	3,269	3,114	2,032	4,871	3,980
9-11 years	4,907	3,645	4,118	2,082	5,503	4,181
12 years	5,580	4,116	5,001	2,292	5,864	4,437
13-15 years	6,084	4,470	5,391	2,293	6,943	4,782
16 and over years	7,633	5,286	6,909	3,357	8,746	4,388

Sources: U.S. Census of Population, 1960: PC(1) 26D, 45D, 51D, Table 138. Spanish surname is for those 25 years and over with income—special report from Bureau of Census, **Mexican-Americans in Southwest Labor Markets,** Walter Fogel, **Mexican-American Study Project**, Division of Research, Graduate School of Business Administration, University of California, Los Angeles, October, 1967, Table 3, p. 12.
   1. White includes Spanish surnames which comprise 52% of the white urban Texas population; Nonwhites have been excluded in computing these medians. Differences would otherwise have been greater.
   2. The nonwhite urban population.
   3. State nonwhite 25 years and above are 97% urban; all nonwhite subtracted from state urban to estimate urban white medians.
   4. State nonwhite essentially the same as urban nonwhite since 88% of nonwhite in Wisconsin are in urban areas.

In the course of his efforts to determine the general relationship of income to education for ethnic groups, Fogel found that between 1950 and 1960 increases in the educational attainment of Spanish surname males in relation to Anglos were greater than increases in their relative incomes. He points out that the experiences of the 1950's suggest that Spanish surname persons could attain educational equality without attaining income equality. He admits, as do others describing the relationship of education to income, that the data on which his conclusions are based are not excellent but that the data do suggest that once a group has been absorbed into the economy within the manual labor sector its income gains will come more slowly than during the period that it is initially seeking entry into the urban industrial economy. We agree that the degree of discrimination in employment affects income despite educational parity of different groups. It is not only that the high educational requirements for some jobs slow progress but it is also the manner in which the labor markets operate—the manner in which non-manual employment barriers may exist through formal educational requirements on one

TABLE 19. EDUCATION AND TOTAL ANNUAL INCOME OF RACINE FAMILIES, 1960 (IN PERCENTAGES)

LEVEL OF EDUCATION

Total Annual Income	No education-4 years			5-7 years			8 years			9-12 years			13 or more years			Not reported		
	M	N	A	M	N	A	M	N	A	M	N	A	M	N	A	M	N	A
$5,500 or more	24	48	14	37	40	58	38	47	43	26	47	68	33	50	77	9	11	0
$4,500–$5,499	15	24	29	13	33	8	24	6	22	23	21	15	33	25	6	22	19	25
$0,000–$4,499	54	24	43	46	20	25	33	31	24	48	25	10	33	13	2	56	47	50
Not reported	7	5	14	4	7	8	5	16	12	3	7	7	0	12	15	13	22	25
	100	101	100	100	100	99	100	100	101	100	100	100	99	100	100	100	99	100

Mexican-American χ^2 = 6.59, 8 d.f., not significant; Anglo χ^2 = 29.81, 8 d.f., p <.001; Negro χ^2 = 11.79, 8 d.f., not significant. Nine to 12 years of education: Mexican-American/Anglo χ^2 = 32.27, 2 d.f., p <.001; Negro/Anglo χ^2 = 16.00, 2 d.f., p <.001; Mexican-American/Negro χ^2 = 6.57, 2 d.f., p <.05. Eight years of education: Mexican-American/Anglo χ^2 = .49, 2 d.f., not significant; Negro/Anglo χ^2 = 4.41, 2 d.f., not significant; Mexican-American/Negro χ^2 = 3.89, 2 d.f., not significant. Less than eight years of education: Mexican-American/Anglo χ^2 = 2.58, 2 d.f., not significant; Negro/Anglo χ^2 = 2.02, 2 d.f., not significant; Mexican-American/Negro χ^2 = 22.32, 2 d.f., p <.001.

hand and informal decisions not to hire Mexican-Americans or Negroes on the other. In Racine, then, large income increases for the Mexican-Americans came when they made the move from migratory labor to industrial labor. Further increases in income resulted as the inmigrant achieved within the industrial hierarchy but gains relative to the total population come only as Mexican-Americans acquire more education and break such barriers to non-manual labor as exist in the larger society. In 1960, Anglos, already the best educated, were increasing their schooling and the need for less educated manual laborers was declining.[17]

In Mississippi, according to Census data, the differences between the urban white and Negro populations of the state were more evident than differences between the urban white and the Spanish surname populations of Texas. Although differences in income related to education were relatively small at the lower educational levels, white income increased more rapidly than Negro income with additional increments of education. Turning to the differences between Mississippi and Wisconsin we examine first the distribution of nonwhite persons. At the lowest educational level Wisconsin nonwhites have a median income two or three times that of nonwhites in Mississippi, and although they are more than $1,000 above their Mississippi counterparts at the highest educational category, income does not vary with education much more than it does in Mississippi. By contrast, the white population of Wisconsin shows far greater variation in median income with education. Again we draw attention to the fact that the unpartialed relationship of education to occupation and income as usually drawn from the Census is misleading because it combines Anglos, Negroes, Mexican-Americans, and other disadvantaged groups, thus making the relationship appear to be high or similar for all, when it is actually lower among Negroes and Mexican-Americans than Anglos.

Having touched upon the pertinent data in Census volumes, let us turn once again to the Racine data. In Racine the relationship of education to total annual income presented a somewhat different picture from the relationship of either education or income to occupation. The 1960 data on income and education (Table 19) reveal that education was significantly related to income for the Anglos but not for Mexican-Americans and Negroes. Within almost every educational level Anglos were skewed toward higher incomes more than Mexican-Americans and Negroes, but the difference was statistically significant only at the 9 to 12 years of education

TABLE 20. OCCUPATIONAL LEVEL OF PRESENT JOB OF RACINE MALES ACCORDING TO RACE OR ETHNICITY, AGE, AND TIME IN THE COMMUNITY, 1960 (IN PERCENTAGES)

	Always lived in Racine		Lived 10 or more years in Racine						Lived less than 10 years in Racine					
	Years of Age		Years of Age						Years of Age					
	35+	35-	35+			35-			35+			35-		
	A	A	M	N	A	M	N	A	M	N	A	M	N	A
Professional, technical, managerial, proprietor, clerical and sales	40	43	2	2	41	0	0	30	0	2	39	2	1	44
Craftsmen, foremen	32	23	20	35	35	14	30	25	17	15	35	7	22	25
Operatives, maintenance and service, industrial laborers	26	34	71	63	22	86	70	45	77	78	21	86	74	32
Agricultural laborers	1	0	7	0	0	0	0	0	7	4	4	5	3	0
	99	100	100	100	98	100	100	100	101	99	99	100	100	101

category. Differences between Mexican-Americans and Negroes were significant at two levels—Negroes with 9 to 12 years of education and with less than 8 years of education had higher incomes than Mexican-Americans.

As we have indicated, the data on income presented something of a problem in that there were more unreported incomes than unreported occupations or educational levels. Twelve of the 70 families for which income data were not available were also missing level of education for the male. Sixty-three people in all did not report level of education. A computation of percentages for "not reported" levels of education showed that education was not reported for 10 percent of the Mexican-Americans, 13 percent of the Negroes, and 1 percent of the Anglos. Nevertheless, if unreported incomes were placed in the lowest educational category, the relationship of education to income shown in Table 19 would remain unchanged.

What is the significance of comparing the data on education and income for the Racine sample with that of Mississippi, Texas, and other Wisconsin data? Income variation has been almost universally accepted as an artifact of differences in opportunities for education and achieved educational levels. We concluded that significant differences in income existed between sample groups in Racine and that they were in part based on differences in education. The size of the samples precluded placing much emphasis on these differences except at the 9 through 12 years of education level where Anglos were clearly higher than Mexican-Americans or Negroes, and at the less than 8 years of education level where Negroes were clearly higher than Mexican-Americans. A doubt therefore arises as to the real weight of education as a determinant of income and the extent to which it is overshadowed by race or ethnicity and other variables.

Occupation and Income and Age and Time in the Community

Occupation and Time in the Community

Additional analyses were conducted holding the variable "time in the community" constant, since there is some tendency for inmigrants to move upward with long residence. The question is whether they were moving upward more rapidly, or less rapidly, in an expanding and increasingly complex economic order than were the Anglos who had always lived in the community. When time in the community was controlled, the occupational level of Anglos who had been in the community 10 years or more was higher than that of Negroes, and that of Negroes was higher than Mexican-Americans, although the only statistically significant difference was that between Anglos and Mexican-Americans. The same relationship existed among those who had been in the community nine years or less.

Income and Time in the Community

When income was related to race or ethnicity with time in the community controlled, the Anglos were significantly higher than Negroes, and the Negroes were significantly higher than the Mexican-Americans, regardless of the time they had been in the community.

Age, Occupation, and Income

Because age is one indicator of years of experience and perhaps of ability to adjust to the requirements of industrial work, those aged 36 or more were compared with those 35 or less. Regardless of age, Anglos were at higher occupational levels than Negroes, and Negroes were at higher occupational levels than Mexican-Americans. Income was analyzed in the same fashion. Among those aged 36 or more, Anglos had higher incomes than Negroes, and Negroes had higher incomes than Mexican-Americans; among those aged 35 or less, Anglos were higher than Negroes and Mexican-Americans, and Negroes were higher than Mexican-Americans but the latter difference was not statistically significant.

Age, Time in the Community, and Occupation

When both time in the community and age were controlled (Table 20), Anglos were at higher occupational levels than Negroes and Negroes at higher occupational levels than Mexican-Americans, regardless of age and time in the community. The differences between Anglos and Mexican-Americans or Negroes were statistically significant at the .001 level in every case but differences between Mexican-Americans and Negroes were not. Older Negroes who had lived less than 10 years in Racine were skewed toward the low end of the occupational level scale while older Negroes who had lived more than 10 years in Racine were mostly skewed towards the high end of the occupational level scale. Young Mexican-Americans who had lived less than 10 years in Racine were mostly skewed toward the low end of the scale while older Mexican-Americans who had lived 10 or more years in Racine were least skewed toward the low end. Younger Anglos and younger Negroes who had lived 10 or more years in Racine were the most comparable in occupational distribution, although still significantly different. Older Negroes and Mexican-Americans who had lived less than 10 years were almost identical in their occupational distribution. Thus far each table has been discussed in terms of variation between racial and ethnic groups on a specific measure. In some instances, as in Table 20, more than one variable has been held constant in order to determine if what seemed to be a racial or ethnic difference would be eliminated by controlling for an underlying variable that would explain the difference.

When reference is made to the existence of a relationship between race and ethnicity and some variable, or the existence of a difference between racial and ethnic groups, it simply means that the relationship is greater than zero. The null hypothesis of no relationship or no difference has been tested. We are, of course, concerned with the strength of these relationships—just how much of the variance was explained by race and ethnicity and how much was explained by age of male or time in the community, or any other variables that were hypothesized to be explanatory. Rather than become involved in presenting measures of the strength of associations, which is not critical at this point, we shall continue to refer to whether or not there were significant differences between Mexican-Americans, Negroes, and Anglos and sub-categories thereof. In a later chapter we shall discuss at considerable length the relationship of father's occupation to male's education to first job to present job, present income, and present level of living, not as a meaningless matrix of correlations showing the relationship of every variable to every other variable, but within an entirely different format in order to show the strength of relationships within a sociologically meaningful, time-oriented, processural framework—an experiential chain. At this point we are simply determining if there was variance where we expected it.

In Table 20 it is shown that occupational level varied with race and ethnicity with both age and time controlled. It was also apparent that regardless of race or ethnicity, occupational level increased with age of male for those in the community 10 years or more, but not

TABLE 21. TOTAL INCOME OF RACINE FAMILIES ACCORDING TO RACE OR ETHNICITY, AGE, AND TIME IN THE COMMUNITY, 1960 (IN PERCENTAGES)

	Always lived in Racine		Lived 10 or more years in Racine						Lived less than 10 years in Racine					
	Years of Age		Years of Age						Years of Age					
	35+	35-	35+			35-			35+			35-		
Income*	A	A	M	N	A	M	N	A	M	N	A	M	N	A
$5,500 or more	73	65	21	58	71	36	45	72	36	52	77	21	34	67
$4,500–$5,499	18	11	19	18	18	25	23	22	3	15	9	24	36	20
$0,000–$4,499	10	24	60	23	11	39	32	6	60	33	14	55	30	13
	101	100	100	99	100	100	100	100	99	100	100	100	100	100

*Not reported excluded from calculations for this table.

for those in the community less than 10 years or for those who had always lived in the community. When the table was approached as though we were primarily interested in the influence of time in the community, with age held constant, it became apparent that occupational level increased with time in the community, although not significantly so, for both age categories of Mexican-Americans and Negroes but not Anglos. Regardless of their age or time in the community, Anglos were always at significantly higher occupational levels than either Mexican-Americans or Negroes—it was difficult indeed for either of the latter to move upward with time in the community.

Age, Time in the Community, and Income

Table 21 presents the data on total family income following the same format as did Table 20 for occupational level. In every age and time in the community category, the Anglo distribution was skewed toward the high income end of the income continuum while the Mexican-American was skewed toward the low end. Incomes of older Negroes who had been in the community 10 years or more were most skewed toward the high end of the continuum and those who were less than 35 and who had lived less than 10 years in Racine were least skewed toward the high end. Younger Anglos who had lived in the community less than 10 years had the highest incomes while younger Anglos who had always lived in Racine had the lowest incomes of any of the Anglo age and time in the community categories.

Although all Anglo age and time in the community categories showed higher incomes than those for Negroes, and significantly higher incomes than Mexican-Americans, the pattern of variation, that is, the rank-ordering of age and time in the community categories for Anglos, Negroes, and Mexican-Americans, was not identical to that for occupations. For example, Anglos, regardless of how long they lived in Racine, had higher but not significantly higher incomes than Negroes but only the *younger* Anglos came *close* to having significantly higher incomes than did younger Negroes. Negroes, particularly those older and living in Racine more than 10 years, had higher incomes than Mexican-Americans but the incomes of younger Negroes who had lived in Racine more than 10 years were not significantly higher.

Perhaps at this point the clearest summation of the analysis that has just been described is that occupation and income vary in a less clearly discernible pattern with age or time in the community than with race and ethnicity. A more comprehensive treatment of income differences will be presented in a later chapter when the antecedent and other variables hypothesized to be explanatory are related to them in sequential chains.

Occupational Level and Income

As we have previously indicated, when occupational level was controlled, Anglos had higher incomes in categories for which there were sufficient cases to compute tests of significance. In the case of industrial laborers, both Anglo and Negro incomes were significantly higher than Mexican-American. Anglos also had significantly higher incomes than Negroes at the level of craftsmen and foremen.[18]

When further controls were introduced, such as occupational level and time in the community, statistically significant differences were found in just two cases: Anglos had significantly higher incomes than Mexican-Americans among industrial laborers in the community 10 years or more, and Negroes had significantly higher incomes than Mexican-Americans among industrial laborers who had been in the community nine years or less.

When occupational level and age were controlled, Anglos had significantly higher incomes than Mexican-Americans among crafts-men and foremen aged 36 or more, and Anglos and Negroes had significantly higher incomes than Mexican-Americans among industrial laborers aged 36 or more. An insufficient number of cases precluded tests of this nature for all occupational levels. Those that were possible indicated that, as time passed, there tended to be fewer differences in income based on ethnicity or race at the lowest occupational levels.

The numbers with which we were working became quite small through successive partitioning (the introduction of controls such as occupational level and age). To preclude ending this chapter with a misleading impression, we shall turn to U.S. Census data for a look at the larger picture. In 1959 the median earnings of automobile mechanics and repairmen aged 18-24 who were nonwhite was $2,114 but for whites it was $2,772. Their earnings increased with age so that in the age 35-44 group, nonwhites had a median of $3,494 and whites a median of $4,962. The difference increased with each five-year age category. Nonwhite carpenters in the age category

18-24 had a median income of $1,733 and white carpenters $3,008. These differentials were found in occupation after occupation, but with less difference in federal or other employment in which an equal pay for equal work policy prevailed.[19]

Summary Statement on Education, Occupation, and Income

Several of the findings described in this chapter should be repeated in summary before turning to the next topic. Although Mexican-Americans and Negroes were initially employed within a limited range of occupations, education was one of the determinants of the level at which they were working. Education as related to differences in occupational level of first jobs was greatest, however, for persons in the Anglo sample. Generally speaking, education was also positively associated with occupational level of present jobs within that range of jobs available to each group, but was statistically significant only for Anglos. But when time in the community was controlled, only those Anglos in Racine 10 years or more had an occupational level significantly related to their stated education.

The general proposition, that a high level of education opens the door to opportunity in the world of work, found only limited verification in the data at hand insofar as it applied to Mexican-Americans and Negroes in a northern industrial city as of the period around 1960. Having a high school education or part thereof did not seem to help the inmigrant minority group member push above the job or occupational level commensurate with an eighth grade education, even with increased time in the community.

It would appear that either the partial or complete high school education available to Anglos was of more economic value to them than was amount or type of education received by Mexican-Americans and Negroes. One must conclude that higher levels of education and longer periods of time in the urban community were associated with higher occupational levels and higher incomes for Anglos in Racine but did not reveal similar associations for the inmigrant Mexican-Americans and Negroes.

Although we have pointed out that Mexican-Americans, Negroes, and Anglos received approximately equal pay for equal work, it must be remembered that education was not the determinant of the work level among these groups to the extent that it was among the Anglos. It gave the Mexican-American and Negro scant satisfaction to know that he could receive equal pay on laboring jobs when he was relatively unable to attain higher level positions similar to those held by Anglos with the same education.

All of this leads us to the question of mobility, intergenerational as well as generational. Norval Glenn has concluded, after careful examination of occupation, income, and educational data, that nonwhite advancement was considerably greater during the 1940's than the 1950's.[20] He further concluded that nonwhites were more nearly equal to whites in education than in occupation. During the 1950's the nonwhite to white ratio for occupations increased by 2.5 percent, the ratio for income of persons increased by 4.9 percent, and the ratio for education increased by 5.8 percent. Despite the problems in this type of comparison, Glenn's data do suggest that at the 1950 rate of change educational equality would come within 60 years; it would take almost 100 years for nonwhites to obtain occupational equality and over 200 years to obtain income equality. This situation would, of course, change if opportunities became commensurate with education.

NOTES

1. Persons are classified as unemployed in the U.S. Census if they are 14 years old and over and not "at work" but looking for work. Examples of looking for work are: (1) registration at a public or private employment office, (2) meeting with or telephoning prospective employers, (3) being on call at a personnel office, at a union hall, or from a nurse's register, or other similar employers, (4) placing or answering advertisements, and (5) writing letters of application. The likelihood that persons at the lowest educational and work experience levels (where Mexican-Americans and Negroes are found disproportionately to Anglos) will be engaging in these activities is considerably less than that for persons at higher educational levels and with more experience in the work force. It does not take much imagination, even for someone who has spent little time in either the Mexican-American or Negro community, to realize that it is easier for an aggressive, experienced Anglo to qualify as unemployed than it is for a Mexican-American or Negro in the southwestern or southern milieu.

2. The unemployment rate for the Racine sample was somewhat lower than the U.S. Census unemployment rate; this may be attributed to the fact that the sample consisted of ever-married persons who have a lower unemployment rate than do single persons.

3. The extent to which one group is underemployed more than another has never been adequately measured. One special Census tabulation has shown, however, that at least as many male heads of families are employed only part-time as are unemployed. Furthermore, overall increases in unemployment

have a greater effect on the jobless rates of blue-collar workers and to the extent that Negroes and Mexican-Americans are concentrated at the lowest levels, they are most affected. See Paul M. Ryscavage, "Impact of Higher Unemployment on Major Labor Force Groups," Monthly Labor Review, March, 1970, Reprint 2664, pp. 21-25.

4. Cotulla had a total unemployment rate of 11.5 percent; separate unemployment figures for Mexican-Americans were unavailable but a good estimate would be between 15 and 20 percent.

5. New Albany had an unemployment rate of 6.4 percent and from this we can estimate that the Negro unemployment rate for the community was close to 15 percent.

6. Any one of a number of cutting points could be selected in order to compare the occupational distribution of groups by race or ethnicity and locale. For example, in Chapter Nine of Grebler, Moore and Guzman, *The Mexican-American People,* New York: The Free Press, 1970, pp. 210-211, tables are presented showing persons in white-collar occupations as a percent of all employed persons. In 1960, 41 percent of the Anglo males in Texas were in white-collar occupations while only 16 percent of the Spanish surname and 9 percent of the Negroes were classified in this fashion. While only 27 percent of the Anglos were in low-skill manual occupations, 61 percent of the Spanish surname and 72 percent of the nonwhites were in this category.

7. Differences in level of formal education attained were presented in some detail in Chapter 4, but not across ethnic lines from one age group to another and from one time period to another. In essence, the record of progress to be found by examination of age differences in schooling was not mentioned. During the period from 1950 to 1960, there was a 37 percent increase in the median years of school completed by Spanish surname persons age 35 and over in Texas, compared to 12 percent for Anglos. Much of this could be attributed to the urbanization of Mexican-Americans in Texas. Likewise, the relationship of education to occupation varies from one age group to another. Age groupings revealed educational progress. The question is whether or not younger Mexican-Americans and Negroes will begin to show the same degree of relationship between education and occupational level as Anglos. An excellent presentation and discussion of the available data may be found in Grebler, Moore and Guzman, op. cit., Chapter 7, pp. 142-150. The entire chapter is recommended for persons with a special interest in the educational institution.

8. A short but cogent discussion of various sources of income data is presented by Herman P. Miller in *Rich Man, Poor Man: A Study of Income Distribution in America,* New York: Thomas Y. Crowell Co., 1965, pp. 18-24, and in a detailed appendix, "The Validity of Income Statistics," pp. 215-237. His analysis and comparison of U.S. Census, Current Population and Internal Revenue Service Survey, and Census Reinterviews Survey data present a convincing argument for our position that the Racine survey data may be compared with U.S. Census data.

9. Differences in family size reverse this rank-ordering of Mexican-Americans and Negroes in the Southwest and in Texas. In 1959 Anglos in Texas had a median per person income of $1,772, nonwhites $775, and Spanish surname $629. In essence, we are saying that the individual Mexican-American is more

likely to improve his position by a move to a northern industrial area than the figures on family income in Table 5 suggest. See Grebler, Moore and Guzman, op. cit., pp. 183-185. Chapter 8 of this volume treats the subject of income differentials in the United States and the Southwest in more detail than does our volume.

In the late 1950's wage rates of migrant labor varied to such an extent by states that the migrant workers could scarcely help but notice the difference. Average earnings per day were $7.94 in Illinois but $5.14 in Texas, according to Daniel H. Pollitt and Selma M. Levine, *The Migrant Farm Worker in America*, a report prepared for the United States Subcommittee on Migratory Labor of the Committee on Labor and Public Welfare, Washington, 1960, p. 54.

10. The Racine data present something of a contrast to the Texas data in this respect. Fogel, in Grebler, Moore and Guzman, op. cit., p. 196, points up the difference for Texas. Spanish surname persons with a median educational attainment of 4.4 years had an average income of $2,241 while nonwhites with a median educational attainment of 7.4 had an average income of $2,146. When adjusted for educational differences, Spanish surname persons had a median income of $3,416 and nonwhites a median income of $2,666.

11. To take an extreme example for comparison, in 1964 nonwhite non-migratory workers in the South were employed an average of 77 days for an average yearly wage of $353. Negroes with employment antecedents such as these did not need to even enter the industrial order to better their position in the North. *The Negroes in the United States: Their Economic and Social Situation*, U.S. Department of Labor, Bulletin No. 1411, June, 1966, Table IIIC-16, p. 170.

In more general terms it should be noted that 540 or 87.1 percent of the 620 counties with the lowest median incomes in 1959 were located in fourteen of sixteen southern states. Sixty-two of the Q-5 counties were located in Mississippi (75 percent of Mississippi's counties), the state ranking lowest on median income and containing the largest percentage of families with 1959 incomes under $3,000. Perhaps even more startling is the fact that 38 counties in Mississippi (61.3 percent of its Q-5 counties) had an increase of 80 percent or more in median family income between 1949 and 1959. Almost any measure of family income still placed Mississippi at the bottom of the 51 states in 1959. See *Family Income and Related Characteristics among Low Income Counties and States*, U.S. Department of Health, Education, and Welfare, Welfare Research Report 1, September, 1964.

12. The poor have been counted many times and in many ways. Although the uniform $3,000 test has been applied more frequently than other measures, it is only one of several arbitrary measures that have been utilized. A variable poverty line which still counts about the same number of poor but delineates a group of different composition is described by Mollie Orshansky in *Counting the Poor: Another Look at the Poverty Profile*. Social Security Bulletin, January, 1965, U.S. Department of Health, Education, and Welfare. See also: Mollie Orshansky, *Who's Who Among the Poor: A Demographic View of Poverty*, Social Security Bulletin, January, 1965, U.S. Department of Health, Education, and Welfare.

But whatever the poverty line, $3,000 is not a very high income considering the fact that the annual cost of a city worker's family budget for an unemployed husband, aged 38, wife not employed outside the home, an eight-year-old girl, and a 13-year-old boy was $6,567 in autumn of 1959 in Chicago, according to the Bureau of Labor Statistics. "This budget was designed to estimate the dollar amount required to maintain such a family at a level of adequate living; according to prevailing standards of what is needed for health, efficiency, the nurture of children, and for participation in social activities," Helen M. Lamale and Margaret S. Stotz, "The Interim City Worker's Family Budget," Monthly Labor Review, August, 1960, U.S. Department of Labor Reprint No. 2346.

13. In 1960 Mexican-American non-casual farm wage workers had average total wage earnings of $1,205 compared with $1,354 for other whites and $777 for nonwhites. Migratory Mexican-American workers had an average wage of $926 per year for 136 days of farm and non-farm work. See Economic, Social and Demographic Characteristics of Spanish-American Wage Workers on U.S. Farms, Agricultural Economic Report No. 27, Economic and Statistical Analysis Division, Economic Research Service, U.S. Department of Agriculture, March, 1963, pp. 10-12. It should also be noted that of the five Southwest states, Spanish surname income per person was lowest in Texas, $629 in 1960. A consequence of the higher median family size for Spanish surname persons was that income per person was lower than that of the nonwhite population, even though the nonwhite population had a higher median family income. A thorough discussion of poverty among Spanish surname persons is contained in Grebler, Moore and Guzman, op. cit., pp. 197-200.

14. Walter Fogel, Education and Income of Mexican-Americans in the Southwest, Mexican-American Study Project Advance Report 1, Division of Research, Graduate School of Business Administration, U.C.L.A., 1965, p. 8.

15. Alan Batchelder not only details why the Negro dollar is second-class money but enumerates other economic considerations that must be made in reference to Negro poverty in "Poverty: The Special Case of the Negro," The American Economic Review, Vol. LV, No. 2, May, 1965, pp. 530-540.

16. This is, of course, more frequently the case with a study such as ours where a relatively small sample of Mexican-Americans, Negroes, and Anglos has been selected for analysis. For the larger population, the United States for example, or even regional segments or major racial classifications, it is possible to develop age-specific figures for education and income. The problem and data describing the relationship of education to income are presented in detail in Chapter VI, "Education and Income," Herman P. Miller, Income Distribution in the United States, U.S. Bureau of the Census, Washington, D. C., 1966.

17. Grebler, Moore and Guzman, op. cit., p. 202.

18. Although it was difficult to make adequate comparisons of Anglos, Negroes, and Mexican-Americans who were on the same jobs in Racine, others have been concerned with this problem and are supportive in their conclusions. In Chapter 10 of Grebler, Moore and Guzman, op. cit., pp. 233-244, Fogel concludes that analysis of occupation and income data supports the inference of discrimination against Mexican-Americans in terms of emphasis on formal schooling as a requisite for high-wage manual employment and in terms of income obtained for given types of employment. Fogel shows that in 1959 Spanish surname males in Texas earned from 61 percent (sales) to 84 percent (laborer) of what Anglos earned on the same jobs. These differences may in part

be explained by jobs depending upon the Mexican-American community and in part to concentration of the Spanish surname population in low-wage areas, such as the Rio Grande Valley, as well as by discrimination.

19. *The Negroes in the United States: Their Economic and Social Conditions,* U.S. Department of Labor Bulletin No. 1511, June, 1966, Table III-A-12, p. 147.

20. Norval D. Glenn, "Some Changes in the Relative Status of American Nonwhites, 1940-1960," Phylon, Vol. XXIV, No. 2, Summer, 1963, pp. 120-122.

Chapter 6

GENERATIONAL AND INTERGENERATIONAL MOBILITY

Mobility and Changes in the Occupational Structure

The past two chapters have presented some notion of the occupational mobility achieved by inmigrant Mexican-Americans and Negroes in Racine in terms of (1) individual occupational movement, and (2) the occupational structure of their communities, counties, and states of origin, as well as the population of the United States or five southwestern states. A similar presentation of median income data for the Racine samples in comparison with median income data for communities and other populations of origin enabled the reader to make a rough estimate of the increases in total family income that were likely to accrue to Mexican-Americans and Negroes as a consequence of their move. Some measure of their relative mobility was also obtained by observing differences between the inmigrants and those remaining in Mississippi and Texas. And in Chapter 4 we found that there was relatively less change in occupational levels between first and present jobs for Mexican-Americans and Negroes than for Anglos.

In order to arrive at a more adequate assessment of the mobility that has taken place, we must examine several time sequences that show the manner in which the occupational or income distributions have changed during the period that the inmigrants were leaving their places of origin and attempting absorption into Racine's economy. The period from 1930 to 1960 is that with which we are most concerned, particularly 1940 to 1960, the time at which our respondents were leaving their places of origin and making their first contacts in Racine.

The comparative position of Mexican-Americans and Negroes at Time 1 (1930 or 1940) in Texas and Mississippi, and inmigrants at Time 2 (1960) in Racine (actually, Mexican-Americans and Negroes in Texas, Mississippi, or Racine in 1960) does not indicate what might have happened to the Mexican-American or Negro inmigrants between Time 1 and Time 2 had they remained in their community of origin or attempted to enter some other urban industrial milieu in the South, Southwest, Far West, or Middle West. Likewise we have barely considered the changes in the occupational structure that have occurred in the South, in the Southwest, in the Midwest, and in the nation as a whole during this period, particularly since 1940 or 1950.[1] Nor have we explored, as we must, the effects of inflation and the rise of salaries in relation to inflation or other causal factors.

Although we have indicated something of the range of possibilities open in 1960 to the inmigrant in the communities from which he came in comparison with the communities to which he went, such as Racine, the stage has not been completely set for a discussion of mobility among Mexican-American and Negro inmigrants to Racine.[2] These data simply inform us what it was like in communities of origin in 1960 and what it was like in Racine in 1960.

In the concluding chapter of this volume we shall present some preliminary data on changes in the occupational and income characteristics of persons in our sample between 1960 and 1970 in comparison with changes in the total economy and changes in the specific areas with which we have been concerned. In brief, what we shall later report is that the occupational upgrading that took place during the 1950's continued at an accelerated rate in the United States, and in Racine as well, during the 1960's. Rapid expansion of the economy and public policy (new and expanded programs for retraining the existing work force and prospective new entrants to the labor force) as well as higher levels of education and a decline in discrimination in hiring made it possible for Negroes to be absorbed into the economy during the 1960-1970 period at a more rapid rate than that of white workers. We shall be concerned with how change in our samples compared with changes in the larger economy during this period.[3]

Occupational Changes in the South and Southwest

Grebler, Moore and Guzman in their volume, *The Mexican-American People,* have concluded that between 1930 and 1960 there has been a very small increase in the relative occupational position of Spanish surname people as compared to Anglos in Texas. Spanish surname males had occupations that generated an Index of Occupational Position amounting to 69 percent of the Anglo index in 1930 and 1950 and it had risen to only 75 percent by 1960.[4] In other words, Spanish surname people had not markedly improved their occupational position relative to that of Anglos during the 20-year period in which Mexican-Americans were leaving Texas and coming to Racine.[5] And it is very clear that it would take relatively little improvement in the position of the average Mexican-American to be better off in Racine than in Texas.[6]

Similarly, Batchelder[7] has pointed out that between 1950 and 1960 the number of manufacturing jobs grew 28 percent in the South but while 12,000 out of 944,000 new jobs went to Negro women, which was proportionally fewer than to white women, none went to Negro men. This change was, of course, taking place during the same period that the mechanization of cotton revolutionized southern agriculture and revolutionized Negro farmers and Negro farm laborers off the farm. Furthermore, Glenn[8] has calculated that during the period between 1940 and 1960 the relative occupational status of nonwhite males as measured by the ratio of the nonwhite to the white index of occupational status, did not improve greatly in any region and that it was, in fact, almost stable in the South. One must conclude that the inmigrant had little to lose and could hardly help but improve himself by a move to the North during the period between 1940 and 1960. Gains in occupational status by nonwhites during this period were through movement into expanding urban industrial occupations that did not result in displacing whites in northern industrial communities.[9] Glenn points out, however, that the gains that took place nationally in the relative occupational status of nonwhites were partially offset by the increased relative unemployment of nonwhite workers.[10]

The Structure of Occupations in the United States:
1940 and 1960 and the Racine Samples

Table 22 has been constructed in order to compare the occupational distribution of persons in the Racine sample with the occupational structure of the United States at two points in time. Two major changes took place in the Anglo segment of the

TABLE 22. DISTRIBUTION OF EMPLOYED U.S. MALES AND RACINE SAMPLE BY MAJOR OCCUPATIONAL GROUP, AND RACE OR ETHNICITY, 1940 AND 1960 (IN PERCENTAGES)

| | Anglo | | | Negro | | | Mexican-American | | |
| | U.S. [1] | | Racine | U.S. [2] | | Racine | U.S. [3] | | Racine |
	1940	1960	1960	1940	1960	1960	1950	1960	1960
Professional, technical, managerial, proprietor	16.5	25.4	28.0	2.5	5.6	1.0	6.5	8.0	1.0
Clerical and sales	13.8	13.7	12.0	2.1	6.0	1.0	6.3	8.0	0.0
Craftsmen, foremen	15.5	19.8	30.0	4.4	8.6	22.0	12.9	16.0	14.0
Operatives	18.8	18.8	16.0	12.2	22.3	33.0	18.8	23.0	26.0
Maintenance and service	6.0	5.5	4.0	15.3	14.4	7.0	6.3	7.0	1.0
Industrial laborers	7.5	6.6	7.0	20.5	27.1	22.0	18.5	14.0	46.0
Farmers and agricultural laborers	20.8	10.1	1.0	41.2	16.1	1.0	29.4	18.0	4.0
Not reported	1.0	0.0	2.0	0.7	0.0	13.0	1.1	5.0	7.0
	99.9	99.9	100.0	98.9	100.1	100.0	99.8	99.0	99.0

Sources: Thomas H. Pattern, Jr., "The Industrial Integration of the Negro," Phylon, Vol. 24, No. 4, Winter, 1963, p. 338; 1960 U.S. Census of Population: PC(2) 1B, Table 6; 1950 U.S. Census of Population: P-E, No. 3C, Table 6.
1. Includes Spanish surname.
2. Nonwhite population of the United States.
3. Spanish surname, five southwestern states.

occupational structure between 1940 and 1960. The first involved shifts out of non-agricultural lower-level positions to positions of a more technical nature that developed at the top of the hierarchy, the professional, technical, managerial, and proprietor categories. The second occurred as more persons moved from agricultural work into other activities. The Anglo sample in Racine had an urban industrial occupational distribution, indicating an opportunity structure in Racine comparable to that of other industrial segments of the larger society.

The Negro segment of the table reveals that even greater changes have taken place throughout the United States in what might be termed the "Negro/white" opportunity structure since some Negro positions are in the white sector of the economy and others are in the Negro community and are dependent upon the latter. While the distribution of Negro occupations does not compare with that for Anglos in terms of achieved occupational levels, the movement out of agriculture was dramatic. The Racine Negro sample is shown to be a population of industrial workers. The total body of data that has been collected for Racine leads one to say that Negroes had been absorbed into the manufacturing sector of the economy and were experiencing the instability that is characteristic of employment of this nature during what should be their stable work careers.[11]

The third segment of the table deals with changes in the position of Mexican-Americans in the total economy. These data are Spanish surname, cover only the five southwestern states, and are for 1950-1960. The movement of Mexican-Americans out of agriculture and into the industrial order has not proceeded at the same rate as it has for Negroes. It is true that the urbanization of Mexican-Americans has resulted in some absorption into the urban industrial economy, but when the matter of relative absorption is considered, the Mexican-American has not kept pace with changes in the over-all transformation of the economy. What the move to Racine has meant in terms of mobility is seen by comparison of the Racine sample of Mexican-Americans with the 1950-1960 occupational distribution in the five southwestern states. Simply stated, the story of Mexican-American mobility is one of movement from agricultural work of the most unstable nature into industrial labor of a relatively unstable nature.

Some of the data in this table which have appeared in the previous chapter are used in a different context here to re-emphasize how the organization of the society (as indicated by the distribution of occupations) is changing. As work becomes reorganized from farm to factory, the employment opportunity structure changes. Those who have previously been fully absorbed into the economy and who have the requisite education and experience are in the best position to capitalize on these changes. And so it is that when we refer to the mobility of the inmigrant we must be concerned about his relative mobility—his ability to move as rapidly as those who have been on the treadmill for many years and many generations.

Income Change and Its Significance

One other measurable fact of mobility should be considered, and that is income. Again, we turn to Grebler, Moore and Guzman, since they have presented an excellent table in their volume based on U.S. Census data for 1950 and 1960.

Grebler, Moore and Guzman (Table 23) reveal that between 1949 and 1959 Spanish surname income increased from 57.2 to 61.6 percent of Anglo income, slightly more than a 4 percent relative increase; nonwhite income increased from 48.9 to 53 percent of Anglo income during the same period, again essentially a 4 percent relative increase. In Texas the relative increase was less than 1 percent for Spanish surname people and there was actually a decline for nonwhites during the 10-year period. In essence, although the Spanish surname person had what amounted to a 48.3 percent

TABLE 23. MEDIAN INCOME OF SPANISH SURNAME PERSONS COMPARED WITH OTHERS, 1949 AND 1959, MALES AND FEMALES COMBINED

	Median Income		Percentage Increase	Percentage of Anglo	
	1949	1959	1949-1959	1949	1959
Southwest					
Anglo	$2,137	$3,351	56.8	—	—
Spanish surname	1,223	2,065	68.8	57.2	61.6
Nonwhite	1,046	1,777	69.9	48.9	53.0
Texas					
Anglo	2,190	3,208	46.5	—	—
Spanish surname	1,134	1,682	48.3	51.8	52.4
Nonwhite	971	1,349	38.9	44.3	42.1

Source: Grebler, Moore, and Guzman, **The Mexican-American People,** New York: The Free Press, 1970, excerpts from page 190, Table 8-6.

increase in dollar income and the nonwhite a 38.9 percent increase, neither was likely to increase his income position relative to Anglos during the period between 1950 and 1960 if he stayed in Texas. There still remains the question of whether the inmigrant's dollar increase in Racine was greater than the dollar increase that he might have had in Texas in terms of purchasing power on the one hand, or in terms of his relative position in the income structure on the other.

This question has been answered, in part, in Herman P. Miller's *Income Distribution in the United States.* The median income of families in constant dollars was $5,660 in 1959 and $3,772 in 1949, but before being transformed to 1959 dollars the 1949 median had been $3,083. In other words, families actually had a median increase in dollars of $1,888 rather than $2,577 during this 10-year period. Although median family incomes for the U.S. increased during this 10-year period, a portion (almost $700, or 27 percent) of this general increase was lost by inflation. Inflation varies from locale to locale but it is safe to say that persons with lower incomes are hit hardest by inflation. Since such a large proportion of the inmigrant income must be spent to merely reach the subsistence level, whatever increase in income is lost to inflation will depreciate the gain in real income that could be utilized in raising their level of living.[12]

Another way to assess the possibility of upward movement in the economy is to look at the percentage share of the total income received by the top 1 percent of the population. In 1913 this sector received approximately 15 percent of the total income of the United States. By 1930 they still received over 14 percent of the total income. During the next 10 years this declined to less than 12 percent but by 1948 only to a little more than 8 percent. Inequality in incomes seems to have declined in more recent years.[13] Miller states that further analyses of the data suggest that the matter is not quite this simple and that very little redistribution of income has really taken place. However, there is one category in which family income has increased—where there are working wives.[14]

What Can Be Concluded About the Mobility of Migrants?

The real question is whether our respondents raised their occupational level and their income more (or relatively more) by moving than they would have by staying in Texas or Mississippi. Given the changes that were taking place in the economy of

Mississippi and the South in general or in the economy of the Southwest (Texas in particular), there was little likelihood that they could have bettered themselves by staying in the communities from which they came. Their best alternative, as they saw it, was to seek economic opportunities in other cities of the South or Southwest, or, as in the case of the Racine respondents, in northern industrial communities. [15]

The disproportional increase in wages and salaries from one major occupational group to another must also be taken into consideration in evaluating the mobility that has taken place within and between generations. For example, farm laborers and farm foremen had a 9 percent decrease in wages between 1950 and 1960 while industrial laborers increased 38 percent, operatives 56 percent, craftsmen and foremen 60 percent.[16] The period from 1930 to 1960 revealed that the directional change from 1950 to 1960 was particularly pronounced for operatives, 325 percent, while least pronounced for farm laborers, 189 percent. If the migrant laborer failed to take the step or steps necessary to move into the urban industrial labor force he was sure to lose in relation to changes in the total economy.

When percentage increases from 1949 to 1959 are considered on a regional basis or statewide basis, the evidence continues to multiply in favor of bettering one's chances by a move to Wisconsin. For example, wages for industrial laborers increased 59 percent in Wisconsin but only 40 percent in Mississippi and 44 percent in Texas. During this same period wages for farm laborers were increasing only slightly or declining in the southern states. Even if the Mexican-American or Negro could have been absorbed into the economy in his home state he would be better off in Wisconsin where the chances were better.[17]

Considering that Racine's most promising opportunities for employment were in metal fabricating industries and that laborers' wages increased 69 percent in Wisconsin in this category between 1949 and 1959, the case for mobility through migration seems even clearer.[18] In machinery manufacturing industries the increase for Wisconsin laborers was 67 percent between 1949 and 1959, and 152 percent between 1939 and 1959.[19]

With these background data placing the Racine study in a better perspective, we may now turn to a measure of intergenerational mobility.

Intergenerational Mobility

Movement from agricultural to industrial labor may or may not be defined as upward—we have considered it upward occupational mobility. If an occupational level cutting point is selected that is sensitive to mobility from agricultural to industrial employment then Mexican-Americans and Negroes would be significantly more mobile than Anglos. But if the cutting point selected is sensitive only to mobility between various industrial positions or between industrial positions and the professions, then the Anglos would be most mobile.[20] We constructed a mobility scale that was sensitive to the mobility of lower occupational level rather than upper level persons. Given the fact that 45 percent of the Mexican-Americans, 43 percent of the Negroes, but only 14 percent of the Anglos had parents who were agricultural laborers or farmers and that 71 percent of the Mexican-Americans and 32 percent of the Negroes but only 6 percent of the Anglo males had done agricultural labor in the past, it was obvious that upward movement at the lowest levels would be registered rather frequently for Mexican-Americans and Negroes in contrast to Anglos. The 1960 intergenerational mobility scale had a low minimum coefficient of reproducibility (.6300) but a coefficient of reproducibility of only .8772, and was therefore not scalable by Guttman standards. It was probably this relative mobility among Mexican-Americans and Negroes which explained their favorable perception of progress since moving to the northern industrial community. Mexican-Americans and Negroes had neither the occupational mobility nor the income increases of the Anglos in Racine, but they had made considerable progress considering their antecedents.[21]

Generational Mobility in the Racine Samples

The occupational mobility of male respondents or spouses of female respondents was measured by a four-item scale (Table 24) in 1960. The male respondents at the highest end of the scale were those who had always been at high occupational levels and who had fathers who worked at relatively high levels. The males at the lowest end of the scale were working at a low occupational level and had fathers who did the same. Three other categories were included: early mobile, intermediate, and late mobile. The respondent de-

TABLE 24. OCCUPATIONAL MOBILITY OF RACINE MALES, 1960 (IN PERCENT-AGES)

Mobility Type	Mexican-American	Anglo	Negro
Constant low status	51	7	32
Late mobile	14	10	26
Intermediate	9	7	16
Early mobile	12	21	16
Constant high status	13	55	10
	99	100	100

Occupational Mobility Scales: [R = .9189; MR = .6150] Mexican-American/Anglo χ^2 = 21.06, 1 d.f., p <.001; Negro/Anglo χ^2 = 19.81, 1 d.f., p <.001; Mexican-American/Negro χ^2 = 2.27, 1 d.f., not significant. Only those respondents interviewed for the first time in 1960 (558) are included in this scale.

scribed as having experienced early mobility had a present job, next to present job, and first job that were at the highest levels on the occupational scale but whose father's position was at the lowest levels. The respondent who experienced late occupational mobility had a present job at the highest level on the scale, but his next to present job and first job had been at the lower end of the scale, the same at which his father worked. The results showed that 55 percent of the Anglos but only 13 percent of the Mexican-Americans and 10 percent of the Negroes were in constant high status. More Negroes than Mexican-Americans were in the early mobile and the intermediate categories while Mexican-Americans were far more frequently in the very lowest category. When the scale scores were dichotomized between late mobile and intermediate types, the Anglos were significantly more mobile than either the Mexican-Americans or the Negroes. The Mexican-Americans and the Negroes were not significantly different. If a lower cutting point had been selected, the Mexican-Americans would have appeared significantly less mobile than the Negroes.

Antecedent Handicap

In an effort to place all of the items presumably indicative of a handicap to either mobility or high level stability in a single index, a five-item scale was constructed from data previously presented. The respondent with the lowest scale score had the lowest potential handicap for adjustment in the urban milieu and the following characteristics: (1) former home was not in southern U.S. or respondent had always lived in Racine; (2) occupation of father of respondent or father of respondent's spouse was professional, technical, managerial, proprietor, clerical, sales, craftsman, foreman, operative, maintenance and service, or industrial laborer; (3) respondent had been in Racine 10 years or more; (4) male respondent or spouse had eight years or more of education; and (5) first job of male respondent or spouse was professional, technical, managerial, proprietor, clerical and sales, craftsman, foreman, operative, maintenance and service, or industrial laborer. The data are presented in Table 25 and showed that Anglos were significantly less handicapped than were either Mexican-Americans or Negroes, and that the differences between the Mexican-Americans and the Negroes were minimal, although Negroes tended to be less handicapped than the Mexican-Americans. Eighty-four percent of the Anglos but only 20 percent of the Negroes and 11 percent of the Mexican-Americans were in the least handicapped category. Twenty-eight percent of the Mexican-Americans, 23 percent of the Negroes, but only 1 percent of the Anglos were in the most handicapped category.[22]

It was risky for the inmigrant to stay in Texas or Mississippi and risky to move to Racine. The changes that were taking place in the economy of Texas and Mississippi were such that whatever the migrant's handicaps as perceived by himself or others, the possibility of success in a high wage industrial community seemed worth the move.

TABLE 25. FIVE-ITEM HANDICAP SCALE (IN PERCENTAGES)

Scale		Mexican-American	Negro	Anglo
Least handicapped	0	11	20	84
	1	24	16	1
	2	6	11	7
	3	6	17	4
	4	26	13	2
Most handicapped	5	28	23	1
		101	100	99

Antecedent Handicap Scales: [R = .8827; MR = .6540] In testing the significance of differences between groups, categories were dichotomized between scores 0-1 and 2-5 for computations of chi square. Mexican-American/Anglo χ^2 = 137.63, 1 d.f., p < .001; Negro/Anglo χ^2 = 137.17, 1 d.f., p < .001; Mexican-American/Negro χ^2 = .15, 1 d.f., not significant.

Some Cases in Point: Mobility as of 1960

Jessie G., a 34-year-old Mexican-American who spoke mostly Spanish, was an example of an inmigrant with multiple handicaps for adjustment in the industrial milieu. He received the highest score on the antecedent handicap scale. Jessie considered Cotulla, Texas, where he resided for 28 years before moving to Racine in 1954, as his former home or hometown. It was in Cotulla that Jessie received his formal education—a total of two years. When friends told Jessie about work opportunities in Racine he went there to find an easier job and one that would pay more money than he was receiving as a ranch hand in the Cotulla area. In his present job he swept up chips from the floors in the factory where he was employed and was classified as an industrial laborer. This position was an improvement over his father's irregular employment on ranches and in migratory labor and his own past position as a ranch hand and sometimes migratory farm worker. It was unlikely that Jessie would be absorbed into the economy at any level other than the lowest.

By way of contrast, Juan D. was born in Racine. Juan received his education through high school in the Racine public schools. Since his father had worked for many years in heavy industry in Illinois and Wisconsin, Juan was more likely to find industrial work than would new arrivals with similar training. His first job had been in heavy industry; since then he had moved upward, acquiring seniority in a position where he was classified as a semi-skilled machine operator. Considering the fact that he was Mexican-American, Juan had done quite well in a community which had traditionally presented limited opportunities to members of his ethnic group.

Jessie, our first example, improved himself by moving from Cotulla to Racine; Juan has been absorbed into the economy at a slightly higher level than his father. But neither Jessie nor Juan would be likely to reach the level of Anglos whose first work was not in heavy industry, or those with considerable education and engaged in atypical first employment—that was at a lower level than that characteristic of the stable work period of most Anglos.

Similar examples could be given for Negroes who had either come to Racine from Mississippi or who had grown up in Racine, but the existence of a sizeable Negro community and the specialization of service occupations within the larger community made it easier for the Negro than the Mexican-American to commence work in Racine anywhere other than in heavy industry. While the Mexican-American is usually forced to compete with the Anglo in the larger community, the Negro inmigrant has the choice of competing with either Negroes or with Anglos in the larger community, or with less effort, to attempt an adjustment within the Negro community.

Many years ago the author conducted a series of experiments with the game "Monopoly" in which each person started with unequal initial advantages representing differential antecedents. More than 100 games were played and it was shown that the outcome of the games was directly related to the initial advantage or disadvantage of the player. Following this experiment, a more complex model was developed in which two Monopoly boards were placed together to form a figure-8 course rather than the basic square course provided by one playing board. One track of the figure-8 represented the larger community and the other the minority group community. Each player could choose to compete with the entire group by going around the figure-8 or could stay on that track of the figure-8 which was representative of the Negro community. Though a disadvantaged person could elect to remain and compete with other disadvantaged persons on his own track, competition with the larger society was not entirely eliminated because at any time the more advantaged persons could elect to leave their track and move down onto the lower board or into the disadvantaged community. And so it is in the real world. The Negro can minimize his contact with the larger Anglo society but only to the extent that the Anglo permits. At any time the Anglos can by one sort of decision or another affect the lives of the persons in the Negro community and force them into a more competitive situation. For example, if an urban renewal program removes a sizeable section of the Negro community, the extent to which removal disrupts the Negro community in turn handicaps the efforts of the Negro in seeking a satisfactory readjustment within his own group.

The importance of the antecedents of inmigrants in adjusting to the urban industrial community has now been mentioned in several chapters. The more inadequate the preparation of the Mexican-Americans and Negroes for competition in the urban industrial milieu the more difficult was their adjustment in the community. But prepared or not, their position in their communities of origin or in the migratory labor cycle was more precarious.[23] Race and ethnicity were efficient predictors of level of economic absorption, but not because there was an innate handicap in being Mexican-

American or Negro, but because (1) the early life experiences of Mexican-Americans and Negroes had handicapped them, and (2) the larger society was organized in such a way as to limit the extent to which they could be absorbed into the economy.[24]

NOTES

1. Until relatively recent years most sociologists were inclined to conclude that social mobility was declining, that the occupational structure was becoming more rigid. The opposite trend for recent years is shown by Otis Dudley Duncan, "The Trend of Occupational Mobility in the United States," American Sociological Review, Vol. 30, No. 4, August, 1965, pp. 491-498, and also by Duncan, "Occupational Trends and Patterns of Net Mobility in the United States," Demography, Vol. 3, No. 1, 1966, pp. 1-18. But at the same time, Duncan, in "Patterns of Occupational Mobility Among Negro Men," Demography, Vol. 5, No. 1, 1968, pp. 11-22, has pointed out that mobility tables for Negro and non-Negro men 25 to 64 years of age indicate quite contrasting patterns. Negro men, regardless of their social origins, were in lower-level manual jobs in 1960 but non-Negroes with favorable occupational origins remained at their father's level or moved upward.

Also see the revised text of Duncan's address as President of the Population Association of America, "Inequality and Opportunity," Population Index, Vol. 35, No. 4, October-December, 1969, pp. 361-366. The reader who has not recently studied St. Clair Drake and Horace R. Cayton, Black Metropolis: A Study of Negro Life in a Northern City, New York: Harper & Bros., 1945; Revised and enlarged edition by Harper & Row, 1962, would do well to look at Chapter 18, "The Measure of the Man," in order to obtain some historical perspective on the occupational structure as it existed for Negroes and whites in the 1930's. In their concluding chapter to the revised edition, "Bronzeville 1961," the authors state that "Even a cursory tour of Bronzeville will reveal that 'The Measure of the Man' remains the same. . . . The changes which have occurred are related primarily to two crucial facts of the 1960's; namely, that America is experiencing a period of prosperity and that Negroes are living in the Era of Integration." ibid., p. xv. Two earlier articles describing trends in the occupational distribution of employed workers since 1910 are Albert J. Reiss, Jr., "Change in the Occupational Structure of the United States: 1910 to 1950," Cities and Society, Paul K. Hatt and Albert J. Reiss, Jr. (eds.), New York: The Free Press, 1956, pp. 424-431, and by Natalie Rogoff, "Recent Trends in Urban Occupational Mobility, ibid., pp. 432-445.

2. There is no better source for a grasp of the subject, crucial variables, and possibilities for measurement than Leo Schnore's review of the empirical and theoretical literature, "Social Mobility in Demographic Perspective," American Sociological Review, Vol. 26, No. 3, June, 1961, pp. 407-423.

3. Changes during the period 1957-1962 are compared with changes during the period from 1962 to 1967 in Claire C. Hodge, "The Negro Job Situation: Has It Improved?" Special Labor Force Report No. 102, Reprint No. 2599,

Monthly Labor Review, January, 1969, pp. 20-28. Hodge concludes that over the past decade there has been improvement in the occupational configuration of Negro employment in terms of pay, status, and security.

4. *Indexes of Occupational Position of Spanish Surname Males and Anglo Males, Texas, 1930, 1950, 1960*

		1930	1950	1960
Spanish surname	(1)	23.8	28.8	33.0
Anglo	(2)	34.7	41.5	43.9
Ratio (1)(2)	(3)	.69	.69	.75

Source: Grebler, Moore and Guzman, *The Mexican-American People,* New York: The Free Press, p. 217.

5. Changes in the employment structure of Texas for the period 1940-1960 have been detailed in an article by Robert T. Newsom, "Metropolitan and Nonmetropolitan Employment Changes: The Case of Texas," Sociological Quarterly, Vol. 50, No. 2, September, 1969, pp. 354-368.

6. Considerable attention was given to the problem of accurately assessing mobility by Walter Fogel, *Mexican-Americans in Southwest Labor Markets,* Advance Report 10, Mexican-American Study Project, Division of Research, Graduate School of Business Administration, UCLA, October, 1967, pp. 79-112. While clear increases in occupational levels were found between the first and second generations of Mexican-Americans in the Southwest, the movement between second and third generations was not so marked. There was less income gain between the generations, explainable in two ways, neither exclusive of the other: some Mexican-Americans who reported employment were unemployed or employed at low wages while others moved up occupationally without income gains, i.e., they moved into low wage professional and managerial jobs. Furthermore, much of the generational upgrading of Mexican-Americans is based on changes in residential location from low wage areas to urban areas paying better wages. R. L. Skrabanek and Avra Rapton describe the changes that have taken place in Bexar County (San Antonio) and Atascosa County (below Bexar) in *Occupational Change among Spanish-Americans,* Texas A & M University, December, 1966. This study, based on Census data and interviews with a sample (268 interviews in Atascosa County and 276 in San Antonio, December, 1961 and January, 1962) of the population, guardedly refers to progress among Spanish Americans, generational and intergenerational, but indicates that these changes must be evaluated in reference to changes in the total economy of Texas and the Southwest. That pattern of occupational change which has taken place is one of decline in farm workers and a corresponding increase in blue-collar workers.

7. Alan Batchelder, "Poverty: The Special Case of the Negro," The American Economic Review, Vol. LV, No. 2, May, 1965, pp. 530-540.

8. Norval D. Glenn, "Some Changes in the Relative Status of American Non-Whites, 1940-1960," Phylon, Vol. XXIV, No. 2, Summer, 1963, pp. 114-115. Also see Leonard Broom and Norval Glenn, Chapter 6, "Occupation and Income," *The Transformation of the Negro American,* New York: Harper & Row, 1965. For a more pessimistic but perhaps realistic interpretation of the

data see: Stanley Lieberson and Glenn V. Fuguitt, "Negro-White Occupational Difference in the Absence of Discrimination," The American Journal of Sociology, Vol. 73, No. 2, September, 1967, pp. 188-200.

9. For an evaluation of the alleged consequences of Negro subordination see Norval D. Glenn, "White Gains from Negro Subordination," Social Problems, Vol. 14, No. 2, Fall, 1966, pp. 159-178.

10. Glenn, op. cit., p. 115.

11. The peculiarity of Racine's economy that makes it possible to have high incomes during prosperity but high unemployment and insecurity during periods of recession has been described in Richard L. Brown, An Economic Base Study of Racine, Bureau of Community Development, University Extension Division and Bureau of Business Research and Service, University of Wisconsin, Madison, 1951. Circa 1950 about 70 percent of those employed in Racine County manufacturing were in durables. Sales of durable goods decline during general business downswings more than sales of non-durables that must continuously be replaced. At that time Racine's economy was not only almost exclusively manufacturing but was largely metal fabricating. And in the case of Racine the fact that such a large component of its manufacturing is farm equipment makes it particularly susceptible to the effects of downswings in business. Furthermore, a large share of Racine's production consists of "producer's goods," and sales of these to other industries are particularly influenced by trends in the economy.

12. Herman P. Miller, Income Distribution in the United States, U.S. Bureau of the Census, U.S. Goverment Printing Office, Washington, D. C., 1966, Table I-7, pp. 16-17.

13. Ibid., Table I-9, p. 19.

14. Ibid., p. 22.

15. This report is not concerned with the nature of migration or migrants; it is concerned with the processes of economic absorption and cultural integration. At the same time, it is necessary to indicate that we are aware of the literature on the sociology of migration that touches on our problem. For example, there is the old question of who migrates. This is not our central concern except insofar as it would play a part in determining the rate at which a group was absorbed into the economy. The quality of migration streams undoubtedly varies with the pushes and pulls of local and metropolitan economics. A few early studies at this point are suggested so that the reader may be aware of the fact that some earlier issues have not been overlooked: W. Parker Mauldin, "Selective Migration from Small Towns," American Sociological Review, Vol. 5, No. 5, October, 1940, pp. 748-758; Gilbert A. Sanford, "Selective Migration in a Rural Alabama Community," American Sociological Review, Vol. 5, No. 5, October, 1940, pp. 759-766. For a more recent view of the social and economic correlates of Negro migration, concluding that economic factors of both a "push" and a "pull" character are the most powerful determinants, see William F. Stinner and Gordon F. DeJong, "Southern Negro Migration: Social and Economic Components of an Ecological Model," Demography, Vol. 6, No. 4, November 1969, pp. 455-471.

16. Miller, op. cit., Table III-6, p. 82.

17. Ibid., Table III-7, pp. 85-86.

18. Ibid., Table V-5, p. 112.

19. Ibid., Table V-6, p. 113.

20. There has been considerable discussion of cutting points in reference to mobility measurement, particularly whether or not movement from farming occupations to urban blue-collar occupations is vertical mobility. Although Lenski and others have expressed doubt about the mobility involved in such a move, we believe that it does represent mobility for the migrant agricultural or other farm worker from the south or Southwest. Gerhard E. Lenski, "Occupational Mobility in the United States," American Sociological Review, Vol. 23, No. 5, October, 1958, pp. 514-523.

21. Measuring occupational mobility along the lines that have been suggested here is only one of many ways of approaching the problem. More frequently the type of table presented in Chapter 4 for first job compared to present job or father's occupation versus son's occupation is utilized. Examples of these approaches abound in the literature with lengthy discussions of research looking from father down to son and son back to father. Several examples should suffice: Richard Centers, "Occupational Mobility of Urban Occupational Strata," American Sociological Review, Vol. 13, No. 2, April, 1948, pp. 197-203; Richard Scudder and C. Arnold Anderson, "Migration and Vertical Occupational Mobility," American Sociological Review, Vol. 19, No. 3, June, 1954, pp. 329-334; Sidney Goldstein, "Migration and Occupational Mobility in Morristown, Pennsylvania," American Sociological Review, Vol. 20, No. 4, August, 1955, pp. 402-408; Elton P. Jackson and Harry J. Crockett, Jr. "Occupational Mobility in the United States: A Point Estimate and Trend Comparison," American Sociological Review, Vol. 29, No. 1, February, 1964, pp. 5-15; Saburo Yasuada, "A Methodological Inquiry into Social Mobility," American Sociological Review, Vol. 29, No. 1, February, 1964, pp. 16-23.

For a more recent discussion of attempts to measure intergenerational mobility see Robert W. Hodge, "Occupational Mobility as a Probability Process," Demography, Vol. 3, No. 1, 1966, pp. 19-34; Judah Matras, "Social Mobility and Social Structure: Some Insights from the Linear Model," American Sociological Review, August, 1967, Vol. 32, No. 4, pp. 608-614; Donald D. McFarland, "Intergenerational Social Mobility as a Markov Process: Including a Time-Stationary Markovian Model that Explains Observed Declines in Mobility Rates Over Time," American Sociological Review, Vol. 35, No. 3, June, 1970, pp. 463-476. The complexity of the problem is dealt with more adequately in McFarland's article than in the bulk of the literature that has appeared.

22. Although this scale is not scalable in the Guttman sense, the coefficient of reproducibility falling below .9000, it is presented here in order to illustrate the extent to which Mexican-Americans and Negroes may be so easily separated from each other by a series of background items. A comparison of military service for each group—61 percent of the Anglo males versus 36 percent of the Mexican-Americans and 40 percent of the Negroes—sheds further light on the antecedents of each group and their chances of work experience or specialized training which might prove useful when seeking employment in the urban industrial milieu.

23. Dorothy Nelkin has described how the nature of migratory agricultural work creates an atmosphere of uncertainty and understandably so in "Unpredictability and Life Style in a Migrant Labor Camp," Social Problems, Vol. 17, No. 4, Spring, 1970, pp. 472-478.

24. Here again the reader is referred to Duncan, op. cit.

Chapter 7

EXTERNAL AND INTERNAL MEASURES OF

ABSORPTION AND INTEGRATION

Introduction

Racine's inmigrant Mexican-American and Negro population have been compared with its predominantly long-resident Anglo population in terms of occupation and income, two external measures indicative of absorption into the economy of the northern industrial community but which do not indicate in full the degree to which they have or have not been integrated into either the larger society or a sub-community. A variety of single and composite measures of level of living including household possessions are presented as additional indicators of economic absorption. Measures of linguistic and other behavior which serve to mark the inmigrant as either oriented toward his prior group identification or in the process of being integrated into the larger society are also presented. These measures are evaluated and in turn compared with the inmigrant's perception of his adjustment in urban society.

Within Racine and its sub-communities as well there were startling differences in the external and internal conditions of homes and in the kind and number of household and other possessions. While a variety of criteria were utilized as indicators of differences in life styles or level of living within and between each of the ethnic or racial samples, we found that the Anglos always ranked highest, the Negroes almost always next, and the Mexican-Americans last. But to evaluate the migrant strictly on his level of living or on behavior that is readily observable by the outsider, such as language usage, is to judge him simply on external criteria; to observe the inmigrant as he sees himself adds an entirely different dimension.

The literature is replete with studies demonstrating that the perception of either social or natural objects takes place within a related context—that the meaning of any event in a person's life depends upon the nature of his background and the context of the event. We would therefore expect Negroes and Mexican-Americans, whose background has been characterized by a whole gamut of deprivations, to perceive life in Racine as a marked improvement over the past. We must not presume, however, that the inmigrant will continue to evaluate himself and his position in the same light at a later date as he saw himself relatively soon after settling in Racine. For that matter, how he looked upon himself from 1959 through 1961 (the period of the study) could differ considerably from how he perceives himself today.

Level of Living

Some Problems in Scale Construction

Before presenting the distribution of material possessions among Mexican-Americans, Anglos, and Negroes, a word should be said about the problem of measuring level of living. There are several approaches that may be made to this problem, each with advantages and disadvantages and none that is entirely satisfactory. Some of the earliest scales were simple additive attempts to place households on a continuum according to the number and kind of possessions to be found therein. The choice of which items should be included and which should not was quite judgmental. A number of techniques for determining the variable rather than equal weight that should be assigned to each item have been detailed in an extensive literature. The subject was presented and thoroughly discussed by Sewell 30 years ago[1] in reference to the problem of measuring socioeconomic status. Though much has been added to the literature, the basic methodological concerns have not changed.

Sewell's basic scale has been shortened, modified, and standardized numerous times since 1940 in order to simplify its administration and to render it appropriate for various populations.[2] Our problem lies in the selection of items for measuring level of living for three different groups, Mexican-American, Negro, and Anglo. Will the items deemed appropriate for each group also measure finer differences within groups as well as between them—essentially the cross-cultural measurement problem?[3] While the items that were selected generated quite different between-groups means, relatively less variation was present within the Mexican-American and Negro samples than in the Anglo.

Selection of items for inclusion in the scales was by Anglo researchers or by Mexican-Americans who had been so integrated into the larger society that they were working with the Anglos and undoubtedly thinking like Anglos. The result was the construction of scales that measured progress in the acquisition of material possessions valued by Anglos. This was satisfactory insofar as devices were needed for measuring integration into the host society but did not take into sufficient consideration the kind of material possessions that Mexican-Americans or Negroes might choose as a consequence of their absorption into the economy and increased purchasing power. The possessions scales that we constructed served the basic purpose for between-groups comparison but three separate possessions scales would have permitted a more precise measurement of the relationship of various determinant variables to differences in level of living within each of the racial and ethnic groups.

One further point should be made in reference to the application of scale analysis to the data. A Guttman scale is simply additive and does not give different weights to each item on a basis of its correlation with some external criteria, for example income or occupation, but it does help to meet the argument that a scale should represent one dimension and have internal consistency. While much has been said about the value of variable weights based on either the correlation of each item with the criterion (item analysis) or the multiple correlation of each item with the criterion (item analysis) or the multiple correlation of items with each other and the criterion (discriminant function), the difference in predictability of scales based on any one of many different weighting systems is not great enough to be of major concern to us.

Inspection and discussion of single variables is usually followed by a Guttman scale based on them. Emphasis has been placed on the 1960 data. Such differences in responses as may be found from year to year generally follow the pattern of variation that would be expected from sample differences previously described.

The Mass Media and Communication

Differential utilization of the telephone, television, and the newspapers among the three samples not only is related to level of living, but moreover suggests differences in the degree to which the various groups have been integrated into the formal communication network of the larger society.

There were significantly more telephones in Anglo than in Mexican-American or Negro homes every year of the study. In 1960, 96 percent of the Anglos, but only 63 percent of the Mexican-Americans and 62 percent of the Negroes had a telephone.[4] There was a television set in practically every home regardless of ethnicity. All of the Anglos, 96 percent of the Mexican-Americans, and 92 percent of the Negroes had a television set, a device which brings into all types of homes a vivid if perhaps unrealistic and somewhat distorted idea of how the more fortunate members of the larger society live and sometimes the less fortunate.[5] The distinctive difference between groups was the greater proportion of *operable* television sets in the Anglo homes.[6] There was a significantly different pattern of newspaper subscriptions based largely on the greater proportion of Mexican-American and Negro non-subscribers; for example, only 5 percent of the Anglos did not subscribe to a newspaper in 1960 in contrast to 18 percent of the Negroes and 28 percent of the Mexican-Americans. The responses were originally coded so as to differentiate newspapers by language and type but there were surprisingly few non-Racine, non-English newspaper subscriptions. There was relatively little change in the proportion of subscriptions from year to year, but some families had dropped their subscriptions and new subscribers had been added.

Possessions

Household possessions such as washing machines, sewing machines, and refrigerators proved to be of variable efficiency as indicators of differences in over-all life style. Although significantly more Anglos possessed washing machines (96 percent) than did Mexican-Americans (91 percent) and significantly more Mexican-Americans than Negroes (83 percent) the real differences were relatively small.[7] The picture in respect to the sewing machine showed greater disparity. Seventy-nine percent of the Anglos, 61 percent of the Mexican-Americans, and 42 percent of the Negroes had a machine; each group had significantly greater ownership than any below it. All Anglos, 98 percent of the Mexican-Americans, and 96 percent of the Negroes had refrigerators. If the year, model, and state of repair of any of the above appliances had been taken into consideration, the differences between groups would have been greater.

The most discriminating possession was the presence or absence of floor covering in the front room; 90 percent of the Anglos but only 31 percent of the Mexican-Americans and 30 percent of the Negroes possessed a fabric rug. Though the percentages quoted are for 1960, they proved to be essentially the same for other years of the study.

Auto ownership was interesting, particularly in the 1959 sample or working class sample where significantly more Mexican-Americans (83 percent) had cars than Anglos (72 percent). In 1960, Anglos more frequently possessed cars (90 percent) and had newer models than the Mexican-Americans who, in turn, possessed (78 percent) cars more frequently than the Negroes (64 percent). The latter comparison could well be related to the structure of the family; Negroes with a larger proportion of families with female heads would tend to possess fewer cars whereas Mexican-Americans, because of their migratory labor background, would be more auto-oriented.[8] In 1961, the significant difference between the Anglos and the Negroes became more apparent. Fifty-four percent of the Anglos possessed a 1957 or newer car while only 24 percent of the Negroes were in this category. Twenty-nine percent of the Negroes as compared to 4 percent of the Anglos were without cars. In 1960 and in 1961 there was a significantly larger number of Anglo families with more than one car than either Mexican-American or Negro.

Fifty-seven percent of the Anglos and 53 percent of the Mexican-Americans owned their homes in 1959. Remembering that the Anglos were drawn from areas contiguous to the Mexican-Americans and were older than the Mexican-Americans, it is not surprising that there was little difference in home ownership; however, considerably more Mexican-Americans than Anglos were still making payments on them. In 1960, 75 percent of the Anglos, 41 percent of the Mexican-Americans, but only 32 percent of the Negroes owned their homes.[9] In 1960, 18 percent of the Anglos were making payments on their houses, 14 percent of the Mexican-Americans, and 12 percent of the Negroes; in 1961, 10 percent of the Negroes and 11 percent of the Anglos were doing so. Thus, of those owning their own homes, proportionately more Negroes were still making payments than Mexican-Americans and proportionately more Mexican-Americans than Anglos. Anglo homes also tended to be somewhat larger than the Negro homes and Negroes had significantly less space per person than Anglos.[10]

Scarcely a study of migration or inmigrants has failed to describe the arrival of new residents to the urban community and the process by which they were directed to that part or section of the community in which others like them had settled. To some inmigrants, the move from railroad station to ghetto was the first step in what could be defined as exploitation of the inmigrant.[11] Sheridan Woods, the "barrio," is one such area in Racine. Though social segregation is touched upon in the last half of this chapter it is more thoroughly examined in a later chapter. Spatial segregation may be measured in numerous ways; discussions of this versus that index have been in the literature for the past 40 years.[12] We hold no brief for one index rather than another but prefer U.S. Census block data (although patterns of spatial segregation may even be found within blocks) rather than Census Tract data.[13]

Taeuber and Taeuber have adopted an index of Negro residential segregation in which "the index value represents the percentage of nonwhites who would have to shift from one block to another to effect an even unsegregated distribution."[14] The median for the 207 cities in the United States for which they had data for 1960 was 87.8. Racine with the same index of 87.8 is exactly in the middle of the distribution,[15] which means that it cannot be considered as highly segregated as other cities but that it is segregated. We may also safely assume that whatever may have been the degree of segregation of Mexican-Americans and Negroes in Racine at the time of the 1959-1961 surveys, it was probably less than in their communities of origin.[16] In the chapter on changes in Racine between 1960 and 1970 we show the extent to which Mexican-Americans and Negroes were clustered at these two points in time.

The Completed Possessions Scales

Several possessions scales were constructed from the above data in order to present differences in levels of living between and within groups more effectively. The scales did permit an increase in predictive efficiency over the modal category of the marginals ranging from 44 to 56 percent. Differences in scale scores between Anglos and either Negroes or Mexican-Americans were statistically

TABLE 26. FAMILY POSSESSIONS SCALES: 1959 AND 1960 (IN PERCENTAGES)

Scale Type	1959 Mexican-American	1959 Anglo
Low 0	7	1
1	15	3
2	15	4
3	32	3
4	21	54
5	4	15
6	6	14
High 7	1	5
	101	99

Scale Type	1960 Mexican-American	1960 Anglo	1960 Negro
Low 0	0	0	3
1	4	0	8
2	17	2	20
3	13	1	21
4	34	7	21
5	23	57	19
High 6	9	32	9
	100	99	101

Scale Type	1960 Mexican-American	1960 Anglo	1960 Negro
Low 0	0	0	2
1	0	0	1
2	3	0	4
3	13	1	6
4	10	2	19
5	17	13	26
6	33	7	25
7	18	45	13
High 8	5	31	5
	99	99	101

7-item scale scores: R = .9160; MR = .7640; Mexican-American/Anglo χ^2 = 129.51, 1 d.f., p <.001.
6-item scale scores: R = .9413; MR = .7600; Mexican-American/Anglo χ^2 = 179.72. 1 d.f., p <.001; Negro/Anglo χ^2 = 218.27, 1 d.f., p <.001; Mexican-American/Negro χ^2 = .89, 1 d.f., not significant.
8-item scale scores: R = .9397; MR = .7600; Mexican-American/Anglo χ^2 = 145.85, 1 d.f., p <.001; Negro/Anglo χ^2 = 193.98, 1 d.f., p <.001; Mexican-American/Negro χ^2 = 2.00, 1 d.f., not significant.

significant and of sufficient magnitude to be considered highly associated with race and ethnicity. The differences between Negroes and Mexican-Americans were not statistically significant. Examples of the 1959 and 1960 scales are presented in Table 26.

The 1959 scale consisted of seven items listed in order most frequently found in respondents' homes or possessed by respondents: refrigerator, telephone, sewing machine, fabric rug in front room, washing machine, clothes dryer, and car model not older than 1957. This scale differentiated the Anglos from Mexican-Americans to an amazing extent, although both samples were residing in contiguous areas. Only 11 percent of the Anglos had a score lower than four, whereas only 33 percent of the Mexican-Americans had a score of four or above.

In 1960, two scales were constructed. The first, designed for comparison with the 1959 scale but omitting clothes dryer, included the following six items commencing with that most often possessed: refrigerator, washing machine, telephone, sewing machine, fabric rug in front room, and 1957 or newer model car. The second scale contained eight items from most frequently to least frequently possessed: refrigerator, television set, washing machine, subscription to newspaper, telephone, sewing machine, fabric rug in front room, and 1957 or newer model car. The six-item scale had a coefficient of reproducibility of .9413 which meant that, knowing a person's score, we could predict with consistency which items were in that person's house with very little error. If a person had a 1957 or newer car he would also possess the remaining items in the scale and his score would be the highest possible, or six. A person with the next highest score (five) would have a fabric rug in the front room and all items except the 1957 or newer model car. The eight-item scale had a coefficient of reproducibility of .9397 and distributed the families over a broader continuum than the six-item scale. For example, if the eight-item scale was dichotomized between scale types 7 and 8, 76 percent of the Anglos would be at the top in comparison with 23 percent of the Mexican-Americans and 18 percent of the Negroes. It is interesting to note that on the 1960 U.S. Census Index of Home Equipment, Racine scored 375 but Union County scored only 320 and LaSalle County 271.[17] These scores are based on all families but if this much difference exists between communities, considerably more is likely to exist for different racial and ethnic segments thereof. Thus, the eight-item scale maximized the difference in the level of living between the Anglo and the Negro or Mexican-Ameri-

TABLE 27. HOME CONDITIONS SCALES: 1959, 1960, AND 1961 (IN PERCENTAGES)

Scale Type	1959		1960			1961	
	Mexican-American	Anglo	Mexican-American	Anglo	Negro	Negro	Anglo
Poor Condition 0	23	12	46	9	39	12	8
1	18	12	8	2	6	20	1
2	11	1	17	4	18	9	3
3	24	5	11	33	7	12	23
Good Condition 4	23	69	19	52	30	47	66
	99	99	101	100	100	100	101

1959 Scale Scores: R = .9150; MR = .605; Mexican-American/Anglo χ^2 = 75.77 1 d.f., p $<$ .001.
1960 Scale Scores: R = .9150; MR = .5775; Mexican-American/Anglo χ^2 = 58.90, 1 d.f., p $<$ .001; Negro/Anglo χ^2 = 26.63, 1 d.f., p $<$ .001; Mexican-American/Negro χ^2 = 8.07, 1 d.f., p $<$ .01.
1961 Scale Scores: R = .9317; MR = .7150; Negro/Anglo χ^2 = 10.85, 1 d.f., p $<$ .001.

[142]

can, but on the other hand, the six-item scale was more responsive to differences within the Mexican-American and Negro groups.

In 1961, two scales were constructed, one with eight and the other with nine items. The eight-item scale included the following from most to least frequently possessed: refrigerator, working television set, washing machine, subscription to newspaper, telephone, sewing machine, fabric floor covering in front room, and 1957 or newer car. The nine-item scale included home ownership or purchasing one, in addition to the items in the eight-item scale and the Negroes were more evenly distributed.

This series of scales demonstrated beyond question that the life styles of the Anglos, Negroes, and Mexican-Americans were significantly different, although none markedly differentiated between Negro and Mexican-American life styles, nor revealed differences within groups to the extent that had been expected.

Home Conditions

Probably no other single factor attracted the attention of professionals and laymen to what they conceived to be the problem of recently arrived inmigrants to Racine as much as the external and internal condition of their homes. Obtaining a reliable measure of home conditions is never easy; having interviewers of diverse backgrounds complicated the problem since their internalized standards make it difficult for them to evaluate home conditions as uniformly as expected (and trained) to do.

In 1959, a comparison of the condition of household (interiors) showed no significant differences, though Anglos were skewed slightly more toward the top of the scale than were the Mexican-Americans. But when the condition of yards and house exteriors were compared, those of the Anglos were rated significantly better than those of Mexican-Americans.

In 1960, remembering that the Anglo sample was representative of the entire community, we were not surprised to find that the Anglos fared significantly better than the Negroes or Mexican-Americans on every item of comparison.[18] Anglo house exteriors were placed in the best category in 76 percent of the cases and the interiors in 52 percent. Negro house exteriors were judged to be in better condition (53 percent versus 37 percent) and Negro households to be cleaner and have better kept furniture (33 percent versus 23 percent) than Mexican-American, with the differences being significant. Although

TABLE 28. HOME OWNERSHIP, BY RACE OR ETHNICITY AND LOCALE, 1960 (IN PERCENTAGES)

	United States		Wisconsin			Texas			Mississippi		
	Total	Urban	State	Urban	Racine	State	Urban	Cotulla[1]	State	Urban	New Albany
Total	62	58	69	63	65	65	64	69	58	57	57
Mexican-American	–	53[2]	–	–	41[3]	–	56[2]	73[4]	–	–	–
Negro	38	36	29	26	33	50	47	–	38	42	43
Anglo[1]	64	61	69	64	66	67	67	65	69	64	62

Sources: U.S. Census of Housing, HC(1) No. 51, Tables 2, 8, 12, 37; No. 45, Tables 2, 9, 25, 40, 41; No. 26, Tables 2, 9, 25, 36; No. 1, Tables 3, 9, 24, 27; No. 4, Table 41; No. 6, Table 41; No. 7, Tables 40, 41; No. 33, Table 41.
    1. Percentage of Anglo home ownership is for whites including Spanish surname, except for Cotulla, which excludes Spanish surname.
    2. Percentage of owner occupied housing of white persons having Spanish surname for Standard Metropolitan Statistical Areas.
    3. Percentage of Mexican-Americans who own their own home based on Racine survey. Survey percentage for Negro was 32%, 75% for Anglo, in 1960.
    4. Percentage is white persons of Spanish surname.

86 percent of the Anglo yards were rated neat appearing and significantly better than those of Negroes and Mexican-Americans, condition of the yards showed no significant differences between the latter; 47 percent of the Negroes kept their yards neat and cut their grass as compared to 43 percent of the Mexican-Americans.

In 1961, the Anglo households and house exteriors and interiors were significantly better than those of the Negro.

The stability of ratings was also interesting from year to year. The data consistently indicated that the Anglos had beautiful yards, well-kept houses, and orderly interiors. Those of the Mexican-Americans were not as well kept as those of the Negroes, although they actually had more possessions. Since the Negroes were paying significantly higher rent or had significantly higher monthly payments on the homes that they were purchasing than the Mexican-Americans, they could be regarded as closer to the Anglo in terms of life style despite having fewer home possessions. A comparison of the type of art found in homes confirmed this appraisal. Though art differed significantly from sample to sample, the art in Negro homes was more like that in Anglo homes.

In order to obtain an overall evaluation of the condition of homes, several items were combined in a scale. The homes with the highest scores had the following characteristics: (1) neat appearing yard with grass cut; (2) house exterior painted, porches painted, in good shape; (3) fabric rug in front room; (4) condition of household neat, clean, and well kept with furniture like new. The data for each year are presented in Table 27; all scales had coefficients of .900 or better. The difference between the marginal reproducibility and the coefficient of reproducibility was as great as or greater than any other set of scales in the study. In 1959, the Mexican-American homes were rather evenly distributed across the continuum of scale scores but only 23 percent of the Mexican-Americans had the highest score possible as contrasted to 69 percent of the Anglos. In 1960, the Negroes and Mexican-Americans were skewed toward the lower end and the Anglos were again skewed toward the high scoring end of the continuum. The difference between each of the groups was statistically significant but Negro and Mexican-American homes did not differ as much from each other as both differed from the Anglo homes. In 1961, the Anglo sample was again skewed toward the highest end of the scale and the Negro sample, although skewed in that direction, was nevertheless significantly different from that of the Anglos. Some of the differences from year to year could be attributed to such factors as a different basis for selection of the samples and the chance of a greater number of homeowners being reinterviewed than persons who did not own homes.

Home Ownership in Racine and Other Locales:
Some Problems of Comparison

The unavailability of home ownership and household possessions data for some areas and locales to which we have referred precludes as complete a comparison for these items as was possible for occupation and income between the Racine samples of Negroes and Mexican-Americans and their counterparts in other sections of the U.S. Table 28 does, however, present sufficient data for purposes of comparison. In every locale except Cotulla, Anglo home ownership was proportionately greater than Mexican-American, a most unusual situation and one apparently inconsistent with every piece of information heretofore presented in this volume. Yet it would not be at all difficult to understand if one had visited this community or looked at the pictures that we took while there. The size and quality of the houses suggest that home ownership did not mean the same thing in Cotulla as it does in a northern industrial community.[19] Neither the size of the lots nor the quality of the housing approached that which would be considered minimal in urban areas of Wisconsin or specifically in Racine. Furthermore, many houses were boarded up, standing empty because the owners had left the community for the North on either a permanent or semi-permanent basis.[20] Those who had remained were apt to own their homes.

The percentage of Negroes who owned their own homes was lowest in every area or locale. Negro ownership was lower in Wisconsin and in Racine than in the United States (total or urban), in Texas, or in Mississippi, particularly New Albany. Here again, one must take into consideration the nature of homes in the South and the likelihood that more of those who remained were likely to be homeowners than were those who had left.[21]

Perhaps even more pertinent in terms of explaining home ownership differences between Racine and places of origin of the inmigrants is that the inmigrants had not been in Racine for a sufficient period of time to enable them to save the funds needed to purchase homes. The fact that 41 percent of the Mexican-Americans and 33 percent of the Negroes were in the process of purchasing homes does suggest that each group was attempting to move in the direction of home ownership in the proportion that characterized home ownership in the locale from which they came.

Financial Behavior

Assistance to others, savings, and installment buying were utilized as indicators of financial behavior. In 1960, 10 percent of the Anglos, 12 percent of the Mexican-Americans, and 17 percent of the Negroes sent money to persons living elsewhere. We did not anticipate that many Anglos would send money to persons living elsewhere, but neither were the Mexican-Americans nor Negroes. There was a significant difference in 1961, however, with 23 percent of the Negroes, but only 11 percent of the Anglos, reporting that they sent money elsewhere.

A question on saving was asked each year of the study; each year there was a significant difference between the Anglos and the Negroes or Mexican-Americans and in 1960 the Negroes were saving significantly more than the Mexican-Americans. In the latter year, 71 percent of the Anglos, 52 percent of the Negroes, and 40 percent of the Mexican-Americans were saving. Considering the income levels of the three samples, it is not surprising that the Anglos were significantly greater savers than the Negroes and Mexican-Americans. It is evident that the inmigrant was a target spender and that the Mexican-American was, as shown by his higher possessions score, more of a target spender than the Negro.

There are undoubtedly differences in the extent to which the level of living enjoyed by a particular group (not necessarily race or ethnic) is paid for in cash or on time installments. Although Mexican-Americans were buying everything on time to a greater extent than were the Anglos in the 1959 sample, the only significant difference between Mexican-Americans and Anglos in 1959 was the greater number of automobiles being purchased on payments by Mexican-Americans.[22] The 1960 sample revealed practically no differences between Anglos and Negroes or Mexican-Americans except more Mexican-Americans and Negroes were buying furniture on time than were Anglos.

The Mexican-Americans generally possessed more appliances and household items than the Negroes, but the latter reported fewer installment purchases than the former. Negroes had better jobs and higher incomes; were paying more on their homes, and saving more than Mexican-Americans. Mexican-Americans did not have as good jobs, had lower incomes, were paying less on their homes and saving less but were putting more into appliances, household furnishings, and automobiles.

Ethnic Practices and Identification

Thus far, the emphasis in this volume has been on economic absorption. As we turn to the broader process of cultural integration, we find that measures of it are not as easy to develop, but as ethnic practices may account for much of the observable differences in life style from group to group, those of the Mexican-Americans were examined at some length.

Linguistic Assimilation

Since language was considered to be a good measure of cultural integration, the following three questions were asked: (1) What language is spoken by you and your wife in your home, Spanish or English? (2) What language is spoken by your children? (3) What language is spoken by people here in Racine that you are closest to? The responses to these questions for each year are presented in Table 29.

In comparing 1959 with 1960, we found no significant difference in response to the first or to the third question. Approximately half the Mexican-American adults spoke only Spanish with their spouses in the home and less than one-half spoke predominantly Spanish or a half-English, half-Spanish combination. Similarly, Spanish or mostly Spanish was spoken by almost half their closest associates in Racine and no more than 30 percent spoke English or mostly English. But 13 percent of the children in 1959 and 9 percent in 1960 spoke entirely or mostly Spanish; 32 percent in 1959 and 26 percent in 1960 spoke a half-and-half combination; and 44 percent in 1959 and 63 percent in 1960 spoke predominantly English. Succinctly, about 10 percent of the Mexican-American parents and 20 percent of their closest friends spoke English entirely but less than 10 percent of the Mexican-American children spoke only Spanish.[23]

Scheff, co-director of the project from 1960 to 1962, has published an excellent paper, "Changes in Public and Private Language among Spanish-Speaking Migrants to an Industrial City," in

TABLE 29. LANGUAGE SPOKEN BY MEXICAN-AMERICANS, 1959 AND 1960 (IN PERCENTAGES)

	Respondent and Spouse [1]		Children [2]		Close Associates [3]	
	1959	1960	1959	1960	1959	1960
Spanish	51	45	8	5	44	33
Mostly Spanish	5	9	5	4	3	14
About half and half	24	17	32	26	26	21
Mostly English	4	7	15	31	6	12
English	13	7	29	32	18	19
Not reported or inapplicable	2	14	12	2	1	0
	99	99	101	100	98	99

Respondent and spouse: χ^2 = 1.38, 1 d.f., not significant.
Children: χ^2 = 9.76, 1 d.f., p < .01.
Close Associates: χ^2 = 1.86, 1 d.f., not significant.
  1. What language is spoken by you and your wife in your home, Spanish or English?
  2. What language is spoken by your children?
  3. What language is spoken by people here in Racine that you are closest to?

which he not only made use of responses to the questions but also incorporated the interviewers' records of language use in the interview.[24] All four items were coded in the same way, "Spanish, mostly Spanish, half-and-half, mostly English, and English." He pointed out that the proportion of respondents speaking English in the four settings was inversely associated with the degree of privacy. Seven percent of the respondents used English in the home, 19 percent used it in conversation with their friends, 32 percent of the children spoke it, and 40 percent of the respondents utilized it in the interview. His first proposition was that culture change takes place more rapidly in the external system than in the internal system. His first corollary was that the host language will be used more in public settings than in private settings which, of course, was found to be true. His second corollary was that over a period of time there will be more change towards the host language in public settings than in private settings. When Scheff analyzed the data for 1959 and 1960 in terms of steps toward English he found that 60 percent of the respondents gave the same answer for both years in terms of home language use, 40 percent for children's language usage, and 34 percent for close friends' language. At the same time, he noted that there was a change toward Spanish over the one-year period of 17 percent for home language, 18 percent for children's language, and 27 percent for friends' language. Although expressing concern over the unreliability of language-usage catagories and the possibility of responses varying a category from year to year without perhaps much change in actual language behavior, he obtained a net proportion of change indicating that there was less change towards English in the home than in either children's usage or usage with close friends. By projecting the differences that he found over a period of years, he suggested that Spanish will be replaced by English among the inmigrants not by changes in language spoken by individuals but by generational changes.

Identification and Customs

The Mexican-American community holds two traditional celebrations each year, May 5 (Batalla de Puebla) and September 16 (Independence). Close to 50 percent of the respondents in 1959 observed Mexican fiestas and/or religious holidays. When questioned about the use of Mexican food approximately 90 percent of the respondents said that they ate Mexican foods. Of these, 12 percent served only Mexican food to the entire family, 49 percent ate Mexican food but did not serve it to their children, and 27 percent served it in their homes sometimes or on special occasions. There were usually male/female differences in responses to questions, particularly those related to the home, but only marked differences were observed in analysis.

When respondents were asked whether they *thought* that the Mexican people in Racine would keep their ways or gradually forget them over the years, 44 percent predicted that the Mexican-Americans would not keep their Mexican ways, whereas 29 percent said they would.[25] However, when asked whether they preferred that Mexican-Americans forget their Spanish customs, 59 percent answered negatively. The following (verbatim) responses illustrate quite clearly:

"I don't hang around but with very few Mexicans. Mexicans will forget their ways, especially the younger generation. I do not want this to happen because I'm proud of my nationality."

"They will forget them. I see the generation of today eating less Mexican food in the home. I would like to see them keep their ways though."

"Well, to tell you the truth, a lot of people who live here for a long time forget. They become American, yes, they will forget. It happened to me. I didn't want it to happen, but lots of us forget anyway. When you reside in a different nation, you have to change. I suppose if I were in China I'd act like the Chinese."

Only 15 percent wanted Mexican customs to be forgotten:

> "The Mexican-American people will forget their ways because the neighbors are an influence on Mexican-Americans. When in America do as Americans. This is the way things should be."

> "Well, the Spanish language they should hold on to and celebrate the 16th of September and the 5th of May. The others [Mexican ways], they might as well forget and live like American people."

> "They will forget. If I have children some day, they will have the way of the North people and in a way that's best."

In answer to the question, "When something happens that might affect the Mexican people here in Racine does the Mexican group stick together or usually not stick together?", 36 percent responded "no" and 19 percent stated that "some do, some don't."

> "They are not united. They are united in bad living; in good living they are not united."

> "I think they are disunited, all the clubs we form fall apart within a short time."

Twenty-nine percent responded in an affirmative manner:

> "Some are united. . . . When they're close friends they try to stop something that goes wrong, help one to each other."

> "Well, to tell you the truth I know there's a club and they are united, but I'm not."

> "I think they stick together. . . . If Mexico and U.S. would have a fight we would all stick together."

This question was followed by: "Generally speaking which Mexican people usually go along with the rest of the group?" Five percent of the respondents made an age-group reference; 10 percent referred to place of origin, e.g., Texas or Mexico; 14 percent mentioned clubs or interest groups and less than 1 percent made a language reference.

The respondents were asked the question "Do some of the Mexican people here in Racine ever do things that irritate you or make you mad?" Forty-four percent gave "no, unqualified" responses. Forty-seven percent replied in the affirmative. When probed for specific grievances, 26 percent mentioned drunkenness and fighting, 16 percent mentioned personal characteristics such as slovenliness and bad manners, and 5 percent mentioned sexual behavior and infidelity. Examples were:

"When they get drunk, especially the young Mexicans. That gives non-Mexicans a bad impression."

"At the dances there are fights and drunken men showing off. That's why we don't go."

"When they do not help on a political campaign. They sit back and let some one else do all the work."

"When they are disrespectful."

"Smoking marijuana or getting into a fight."

"Things like going downtown in their dirty clothes."

"Clubs don't help us, they take our money but don't help us."

"When they get drunk and when they become abusive. It is on this that they criticize us Mexicans."

A question arose in our minds as to the emphasis northern urban industrial Mexican-Americans placed on *compadrazgo* and the part family ties played in the cultural integration of their group. During the pretesting period in Sterling various members of the research group spent a great deal of time informally with members of the Mexican-American community and observed younger persons continuously addressing older persons when appropriate as "compadre" or "comadre." Later they were informed by members of the Mexican-American group that reference to compadres had increased after the research group had entered the community in order to conform to the behavior pattern expected of them. Such reaction led us to question the validity of research in which the researcher is defined by the subjects of the control group as a person who expects the group to react in predefined fashion. Yet when we approached the Mexican-Americans for assistance in securing interviewers and other aides, they did tend to turn to members of their immediate family or to compadres, an action which seemed to verify the existence of close Mexican-American familial ties. Though we had never attempted to relate compadrazgo to work in a formal fashion or checked upon the extent to which compadres have managed to assist each other economically or politically in the Mexican-American community, our general impression was that family ties among the Mexican-Americans were much stronger than in the corresponding Anglo community. One businessman told us that he had sold furniture to the senior member of a very large family many years before and that, as a consequence, almost every member of the family purchased from the store. The response to the question, "In your family, do you feel that your compadres are like members of the family?" was most revealing; 77 percent of the Mexican-American control group answered in the affirmative.[26]

Decision-Making in the Home

According to the traditional picture of family structure and decision-making, each sample should have differed in its response to questions on the role of the husband and wife in certain categories of decisions.[27] But perhaps the matter is more complex. Differences in family structure and family decision-making that are assumed related to race or ethnicity might prove non-existent in samples of equal socioeconomic status.

Each year, respondents were asked the following question: "When a large, expensive item is bought, who decides to buy it? You or your wife?" Although in 1959 the husband made the decisions alone more frequently among the Mexican-Americans (19 percent) than the Anglos (13 percent), the difference was not statistically significant. Was this because the Mexican-Americans and Anglos were not too different in socioeconomic status? In 1960, the Anglos responded that decisions were made jointly in 71 percent of the cases and that the husband made the decisions alone in 12 percent; by contrast, 55 percent of the Mexican-Americans made joint decisions while 20 percent responded that decisions were made by the husband alone. Seven percent of the Anglos and 11 percent of the Mexican-Americans stated that the decision was made by the wife alone. Because the families in the Negro sample had a female head more frequently than others, the 44 percent joint response of Negroes and 17 percent wife alone response appeared reasonable. The response pattern of Anglos in 1961 was almost identical to that of 1960 Anglos. While the response pattern of Negroes remained significantly different from the Anglo pattern in 1961, there were more wife alone (20 percent) or joint decisions (53 percent) than husband alone (13 percent) in 1961 than in the previous year. This difference could, of course, simply be due to the somewhat higher socioeconomic status of the reinterviewed group of Negroes in 1961.

How the Respondents See Themselves

Reason for Migration

In order to understand how inmigrants perceive their own

adjustment it is necessary to ask them why they left their former home and came to Racine. Mexican-Americans and Negroes were more likely to have left for job-oriented reasons than were Anglos. In the 1959 sample, 70 percent of the Mexican-American males as contrasted to 43 percent of the Anglo males had come to Racine for predominantly work-oriented reasons; 15 percent of the Mexican-Americans but only 8 percent of the Anglos came to obtain a specific job. Predominantly family-oriented reasons brought 49 percent of the Anglo males. Female respondents answered in much the same fashion with 58 percent of the Anglo women as compared to 25 percent of the Mexican-American females coming for predominantly family-oriented reasons. In 1960, 85 percent of the Mexican-Americans, 68 percent of the Negroes, and 47 percent of the Anglos came for job-oriented reasons. The difference between Mexican-Americans and Anglos was significant both years; the differences among Anglos, Negroes, and Mexican-Americans were significant in 1960.

A Mexican-American who considered his home to be San Antonio, Texas, stated that he had left because:

"There we couldn't earn much money. The jobs paid very cheap, 35 to 40 cents per hour. Even the factories there pay low wages, and the cost of living isn't that different. Things are getting better there, but we still earn more here."

A Negro male left his home in New Albany, Mississippi:

". . . because I was looking for work. There was no work in New Albany. New industry was coming into New Albany but no Negroes were being hired. They might hire one, let him work a while, and then lay him off and hire another, but never for more than a week or so. Negroes were never even hired for foundry jobs."

Anglos who had come to Racine responded in this fashion:

". . . Father lived here and owned a home and was renting it out. We came here so he wouldn't lose it."

". . . my folks decided that, not me. My brother was in the Navy and was stationed at Great Lakes. He liked it so well that the family decided to move here."

The authors made a number of trips in 1965 and 1966 to communities of origin in Mississippi and Texas and spoke at length with Negro, Mexican-American, and Anglo residents about the migration process and the decision to move. Again and again the reasons for migration given by the Negroes and Mexican-Americans were confirmed by Anglos in responsible positions in the community. In Mississippi the Anglos spoke of the lack of opportunity for all races in certain parts of that state. Though they did not say that there were no industrial opportunities whatsoever for Negroes, even as late as the mid-1960's, they emphasized that the Negroes were not hired in industry in the communities where they resided.

Respondents were asked the question, "Why did you come to Racine rather than to some other place?" Thirty-seven percent of the Anglos but only 10 percent of the Mexican-Americans and 5 percent of the Negroes came to Racine with a specific job in mind; an additional 44 percent of the Mexican-Americans and 45 percent of the Negroes but only 12 percent of the Anglos had *heard* of employment opportunities before coming.

People leave their home for job-oriented reasons but do not always pick a destination because the best job opportunities are there. Rather, they have kin in that place, who may or may not have informed them about work opportunities prior to their arrival. Only about 25 percent of Mexican-American respondents had planned to move to a place other than Racine when they left Mexico or Texas. Further questioning revealed that over half the Mexican-American respondents had heard about Racine from relatives living there and that another third had heard about the city from friends or from both friends and relatives. The experiences of the author in Cotulla, Texas, confirmed the findings. After entering a local restaurant and placing an order, the waitress was asked if she had ever heard of Racine, Wisconsin. She immediately smiled and responded in the affirmative. In a few minutes she returned from the kitchen with the cook and soon after brought another waitress over to the table; both were familiar with Racine. In Pearsall, Texas, the author got out of his car, stopped the first person on the street, and asked him if he had ever heard of Racine or of anyone who had ever worked there. He was immediately instructed to go to the yellow house half a block down the street to talk to members of such a family. Numerous examples of similar replies could be given.

In 1961, when Negroes were asked if discrimination was one of the reasons why they left their hometown, about 37 percent of the Negro males and 21 percent of the Negro females responded in the affirmative. The Negro, whose hometown was New Albany, Mississippi, stated that he left because of such incidents as the following:

"While walking on the street you may be called all kinds of names. You may have to get off the sidewalk if two or three were walking along.."

Another Negro complained:

> "We couldn't get a decent job like the white men, and there were certain
> places we couldn't go there like cafeterias."

If the respondent answered that discrimination was not his reason for leaving his hometown, he was asked whether he was treated differently in his hometown because he was a Negro; an additional 7 percent indicated that this was so. One respondent said:

> "Well, we always were served last in stores, and they made you say, 'Yes,
> sir,' and 'No, sir.' "

It is interesting to note that some of those who responded that discrimination was not the reason for their leaving their southern home had added that if such were the case they would not have remained in Racine.

Perception of Change

In 1959 the following question was asked of Mexican-American and Anglo respondents: "Sometimes you hear people talking about the way life has changed by coming to Racine. From what you have seen, what kinds of big changes take place in the lives of Mexican people when they come to Racine?"[28] Sixty-seven percent of the Mexican-Americans responded that changes were good or mainly good; only 13 percent of the Anglos agreed whereas 84 percent did not know or had not perceived any changes in the Mexican-Americans. Since the Anglo might have known little or nothing about the life of the Mexican-American prior to coming to Racine, he perhaps was incapable of making a comparison, but typical Mexican-American responses were as follows:

> "Life in Racine is more comfortable. The people that come to Racine feel
> more at ease, feel the people treat them as people. The environment,
> neighbors give people the desire to keep up to them. They pick up new habits
> in sanitation and keeping up their house."

> "I can't find anything except that I am happy. Everyone has a job and I
> am happy and everyone is better off. And most of all I like the climate and
> the place itself. There's no distinctions here. We're treated equal."

When Mexican-Americans were asked what kinds of changes they perceived among themselves as a group, 59 percent mentioned economic improvement while only 9 percent of the Anglos perceived the Mexican-Americans as having benefited in this manner. When a scale of perception of change for Mexican-American people was constructed with the respondent who scored the highest reporting overall satisfaction, economic improvement, better opportunities, and improved social relations, significantly more inmigrant Mexican-Americans than Anglos reported that life had changed for the better. In fact, when the scale was dichotomized at a point producing maximum discrimination between the two distributions, 58 percent of the Mexican-Americans as opposed to 10 percent of the Anglos were in the category of perceiving favorable change for Mexican-Americans.[29] Mexican-Americans tended to refer to Mexicans in general rather than to themselves individually when speaking of the benefits accruing by moving from a former place of residence to Racine. A typical response was as follows:

> "No, my life hasn't changed. I have always been happy. I have always been the same . . . hard working over there and hard working here. I am older, yes."

> "I would say that our living standards are about the same as they were in Texas. There isn't too much difference. The jobs are hard but they pay better."

When each was asked how his or her life had changed overall after coming to Racine, 66 percent of the Mexican-Americans and 80 percent of the Anglos answered favorably. On a scale similar to that for group perception Anglos placed themselves more toward the favorable end of the scale than did Mexican-Americans but the difference was not statistically significant.[30]

In 1960 the respondents were asked: "Overall, how do you feel about this community when compared to (former home)—more satisfied, less satisfied, or about the same?" Seventy-nine percent of the Mexican-Americans, 74 percent of the Negroes, and 70 percent of the Anglos were more satisfied with the community, with or without reservations.[31] Only 7 percent of the Mexican-Americans, 14 percent of the Negroes, and 13 percent of the Anglos were less satisfied. Though the differences between the groups were not statistically significant, when the question was coded in terms of jobs in the community as compared with those in former home, the differences were statistically significant: 91 percent of the Mexican-Americans, 85 percent of the Negroes, and 74 percent of the Anglos thought jobs were better. All of the groups were generally satisfied with changes that had occurred as a consequence of moving, but why were the Mexican-Americans and Negroes either a bit more satisfied than or as equally satisfied as the Anglos who were doing considerably better or who had made a more satisfactory adjustment in the community?

Comparison of Perception of Change, 1959, 1960

The 1959 and 1960 responses to questions concerning changes brought about by a move to Racine were recoded and presented in Table 30 so as to make them comparable within and between ethnic groups. The 1959 sample of inmigrant Mexican-Americans evaluated the changes that had taken place in their lives as somewhat less satisfactory (73 percent good or mainly good) than did the inmigrant Anglos (78 percent good or mainly good) but the difference was not statistically significant. In 1960, the Anglo and Mexican-American evaluations were again not significantly different, 78 percent of the Anglos and 79 percent of the Mexican-Americans being in the good or mainly good categories. The difference in percentages between the two groups from year to year may partly be due to the differences in

TABLE 30. COMPARISON OF RESPONSES ON CHANGES BROUGHT ABOUT BY MOVE TO RACINE, INMIGRANTS ONLY, 1959 AND 1960 (IN PERCENTAGES)

Global Changes	Mexican-American			Anglo		
	Total	Male	Female	Total	Male	Female
1959						
Changes good	43	48	38	62	70	53
Changes mainly good	30	27	33	16	15	18
Changes half and						
half or same	15	15	16	14	10	18
Changes mainly not good	9	9	9	3	5	0
Changes not good	3	2	4	5	0	12
	100	101	100	100	100	101
1960						
Changes good	32	33	32	15	14	15
Changes mainly good	47	41	52	63	52	75
Changes half and						
half or same	16	19	12	12	24	0
Changes mainly not good	5	6	4	10	10	10
Changes not good	1	2	0	0	0	0
	101	101	100	100	100	100

1959: Mexican-American/Anglo χ^2 = 3.74, 1 d.f., not significant.
1960: Mexican-American/Anglo χ^2 = .01, 1 d.f., not significant.

composition of the Anglo sample in 1959 and in 1960. On the other hand, when the two categories, "changes good," and "changes mainly good," are added, their sum remains the same for the Anglos for 1959 and 1960, namely 78 percent. Greater variation between 1959 and 1960 appears when the males or females of either group are compared with each other. Again, opinions of the Anglo females vary to a far greater degree than do those observed for the Mexican-American females. [32]

The stability of responses to evaluative questions has often been raised. Thirty-two percent of the Mexican-Americans as contrasted to 15 percent of the Anglos changed their response in a positive direction between 1959 and 1960; 32 percent of the Mexican-Americans as contrasted to 51 percent of the Anglos had a more negative response. On the other hand, we contend that the 1960 Anglo sample was less likely than the Mexican-American to have a positive evaluation of changes in way of life brought about by moving to Racine because the Anglo had not experienced the same difference in degree of upward occupational mobility or improvement of level of living as the Mexican-Americans.

Comparison of former home with Racine as a place of residence in the future gives further indication of inmigrant satisfaction with Racine. Although half the Mexican-Americans in the 1959 sample indicated that they had visited their former home, and the same percentage was obtained in 1960, only about 25 percent expected to return permanently to their point of origin. Although 80 percent of them stated that they had visited their hometown, 60 percent said they definitely had no plans to return permanently and only 20 percent indicated that they did have plans to return permanently.

Perception of Discrimination

In 1961, Negroes and Anglos were asked a series of questions about perception of equality of opportunity for people in Racine, the responses to which are shown in Table 31. Negroes did not believe that there was equal treatment for all people in Racine and their answers were significantly different from those of Anglos on almost every question. The most apparent differences between the two groups were in their perception of differential service in restaurants and job opportunities. Mexican-Americans and Negroes differed to a lesser degree in their responses to questions about the equality of treatment in schools, housing, transportation, and taverns

TABLE 31. NEGRO AND ANGLO PERCEPTIONS OF EQUALITY OF OPPORTUNITY IN RACINE, 1961 (IN PERCENTAGES)

	Housing		Restaurants		Taverns		Jobs		Movie Theaters		Transportation		Schools	
	N	A	N	A	N	A	N	A	N	A	N	A	N	A
Strongly agree	2	3	1	3	1	1	2	1	4	16	4	14	4	16
Agree	4	14	14	41	7	16	7	25	77	71	80	84	72	78
Pro-Con	0	6	3	10	6	8	2	12	4	3	5	1	7	2
Disagree	46	59	50	34	53	51	34	45	3	3	5	0	11	3
Strongly disagree	47	14	29	3	24	7	54	11	0	0	0	0	2	0
Not reported[1]	0	4	4	10	9	17	2	6	12	6	5	2	4	1
	99	100	101	101	100	100	101	100	100	99	99	101	100	100

Housing $\chi^2 = 15.61$, 1 d.f., p $<$.001.
Restaurants $\chi^2 = 49.05$, 1 d.f., p $<$.001.
Taverns $\chi^2 = 8.25$, 1 d.f., p $<$.01.
Jobs χ^2 33.73, 1 d.f., p $<$.001.
Movie Theaters $\chi^2 = .00$, 1 d.f., not significant.
Transportation $\chi^2 = 15.53$, 1 d.f., p $<$.001.
Schools $\chi^2 = 16.28$, 1 d.f., p $<$.001.
1. "Not reported" categories eliminated in calculating χ^2.

in the community. Only on equality of treatment in movie theaters was there essentially no difference between the Anglo and Negro responses.

A scale measuring perception of discrimination was constructed from these items. The respondent who perceived the least amount of discrimination agreed that there was equality of treatment in Racine transportation, schools, movie theaters, restaurants, jobs, taverns, and housing, in that order. When the scale was dichotomized at midpoint, the Negroes were skewed toward the end of the scale indicating perception of discrimination (82 percent of the Negroes versus 46 percent of the Anglos).[33]

Suggestions for Change

In the 1961 survey Negro and Anglo respondents were asked what changes or improvements they desired for Racine. Their responses are presented in Table 32. Twenty-nine percent of the Negroes said they would like better housing [34] for Negroes and 39 percent said they would like more and better jobs for Negroes.[35] The most frequent response among the Anglos was that of the 15 percent who said they would like more industry in Racine, more jobs, and would not want industry to leave the city. Another 12 percent indicated that they desired better municipal services, such as better buses, garbarge collection, mail, sewers, water, harbors, and schools. Eleven percent of the Anglos wanted a better downtown area and better shopping facilities; another 10 percent chose more recreational facilities. What is interesting was that none of the Anglos mentioned better housing for Negroes, better and more jobs for Negroes, or better social conditions for Negroes. In essence, there were two separate lists of changes or improvements for Racine, one from the Negro and the other from the Anglo point of view.

When questioned specifically about better housing Negroes stated that rents were too high, many houses were dirty, they were restricted to undesirable areas, they were more crowded than others, and it was difficult to find a place to live in the community:

"Living conditions . . . You might as well come down and say it. Well, I'll tell you. It's not too much different between here and the South. The people in the South come out clear with theirs; here they hide from it, or try to hide it."

"As far as housing, I would like to see it made a little better for the Negro to *buy* a place; renting too comes under this. I believe we are getting the low end of the deal here . . . I believe that as far as getting a home, he [the Negro] has a harder time with a down payment. As for parts of town, they live all over the town, but it's segregated in a way because of the amount of money you have to pay to get it. It doesn't bother me, but you should be able to live where you choose."

"For one thing, the working man's money leaves him fast. House rent is too high and utilities are too high; wages are gone. Places rent from $80 to $125 a month for colored people and the places are rundown. They need rent control here. This is the main problem. The working situation has improved this year. The whites seem to be a little scary this year; I guess because of the business down South . . . The commode leaks and most of these houses are full of roaches and rats, and we have to get rid of these pests ourselves. Aside from the high rent, we have to pay for water. My husband has to fix things around here."

TABLE 32. SUGGESTIONS FOR CHANGE AND IMPROVEMENT IN RACINE, 1961 (IN PERCENTAGES)

	Negro			Anglo		
	Total	Male	Female	Total	Male	Female
Better and/or more jobs for Negroes	39	48	33	0	0	0
Better housing for Negroes	29	20	36	0	0	0
Better social conditions for Negroes	7	10	4	0	0	0
More industry in Racine; more jobs; keep industry from moving out	5	3	6	15	18	13
Municipal services—buses, garbage, mail, sewers, water, harbor, schools, etc.	2	2	1	12	16	7
Better downtown area; shopping, better upkeep	0	0	0	11	7	15
More recreational facilities	3	0	5	10	4	15
Better streets or parking	2	3	1	9	8	9
Taxes lowered	0	0	0	8	14	2
Criticisms of municipal government	0	0	0	4	5	3
Other	4	5	3	8	8	7
Inapplicable; no changes desired	10	9	10	24	19	29
	101	100	99	101	99	100

Negro/Anglo χ^2 = 187.78, 1 d.f., p $<$ .001.

When asked about their concern over jobs Negroes stated that they were either unable to find employment or could only get the poorest jobs. They added that they found it difficult to secure office or clerical jobs and were unable to make use of the education which they had.

"Very few of our people have these good jobs, i.e., department stores, supermarkets . . . We go to school and get our education like the rest of them, but we can't put it to any use."

"Colored should be working in some of these grocery stores and dry goods stores. Even in the foundries there are no colored bosses . . . I believe that there would be better opportunities for the colored race if they had colored bosses in these dry goods stores. Then maybe colored could get other jobs in these places."

"Lots of places won't hire colored. Well just about four or five places will hire colored. I'm laid off now. If these other places would hire colored I might be working now."

By way of contrast, the Anglo answers expressing concern about recreational facilities, better parking, the upkeep of the downtown area, taxes, municipal services, and municipal government ran in this vein.

". . . improve the streets, they're terrible. Racine has the worst streets of any town I have ever lived in. The side streets are particularly rough and full of holes."

"The people don't get together or do anything or support anything for the teenagers. There is more stuff for the older people than the teenagers. I have teenagers and there is nothing for them to do . . . The kids wouldn't be running on the streets and getting into trouble if they had more to do. The Y is too high for poor families with kids."

". . . house taxes are too high. I think they are unfair; they go according to income rather than property."

". . . a new mayor, a new council, and city management, that's what we've got to have. Policemen need higher wages, too."

"I'd like the downtown district improved. Those rotten buildings are so bad, you can't use the second floors of many. Downtown is everything; you judge a town by the way a downtown looks. If there's nothing downtown you feel there is nothing."

Since the question was presented in an open-ended fashion we believed that a different evaluation of the community would be obtained if respondents were confronted with a forced choice type of question (see Table 33), that is, requested to give their views on

TABLE 33. NEGRO AND ANGLO EVALUATIONS OF RACINE AND ITS INSTITUTIONS, 1961 (IN PERCENTAGES)

	MALES				FEMALES			
	Very good	Good	Not very good	Not reported	Very good	Good	Not very good	Not reported
Chances to get ahead								
Negro	3	12	78	7	3	13	65	19
Anglo	6	57	32	5	10	50	29	12
Race relations								
Negro	0	37	59	4	14	44	33	9
Anglo	14	67	15	4	11	68	17	4
For raising a family								
Negro	17	49	31	3	15	45	32	8
Anglo	32	62	5	1	32	63	4	1
Job opportunities								
Negro	2	9	87	3	0	4	87	9
Anglo	3	32	57	8	5	29	54	12
Entertainment								
Negro	2	24	70	5	4	15	65	16
Anglo	8	37	53	2	6	35	55	3
Street construction and maintenance								
Negro	7	37	56	0	14	36	44	6
Anglo	6	55	38	1	4	55	37	3
Parks and playgrounds								
Negro	24	48	24	5	32	36	23	9
Anglo	34	51	15	1	27	52	16	4
Labor relations								
Negro	5	29	51	15	6	33	36	24
Anglo	5	50	39	6	2	44	35	19
Friendliness of the people								
Negro	12	63	20	5	23	55	19	2
Anglo	22	64	14	0	30	46	25	0
Shopping								
Negro	29	39	20	12	23	49	28	0
Anglo	18	60	19	3	16	55	28	1
Public schools								
Negro	29	60	7	5	27	55	14	4
Anglo	34	50	12	5	26	56	11	7
Downtown parking								
Negro	9	42	44	5	10	37	37	15
Anglo	4	26	63	6	6	36	51	6

specific subjects such as chances to get ahead, race relations, raising a family, job opportunities, entertainment, street construction and maintenance, parks and playgrounds, labor relations, friendliness of people, shopping, public schools, and downtown parking.

Seventy-eight percent of the Negro males thought that the chances to get ahead in Racine were not very good while only 32 percent of the Anglo males agreed. Fifty-nine percent of the Negroes did not think that race relations in Racine were very good but 81 percent of the Anglos thought that they were good or very good. Significantly more of the Negroes than Anglos did not think that Racine was a good place for raising a family.

When job opportunities were mentioned, 87 percent of the Negroes said they were not very good while only 56 percent of the Anglos responded in this manner, another statistically significant difference. Negroes did not differ as much from Anglos in their evaluation of entertainment facilities, but rated them considerably lower than did the latter.

That 55 percent of the Negro males said that street construction was not very good but 56 percent of the Anglos said it was good suggests a different focus of attention on the part of the respondents or varying degrees of maintenance in different parts of the community. There was no great difference in evaluation of parks and playgrounds.[36] Fifty percent of the Negro male sample but only 29 percent of the Anglos indicated that labor relations were not good in Racine. Although there was less difference than on items previously mentioned, Anglos thought that the people in Racine were friendlier than did the Negroes. More Negroes thought that shopping was "very good" than did Anglos. One need only envisage their probable reference point to understand a more favorable evaluation on this item.[37] Negroes and Anglos differed little in their evaluation of the public schools.[38] Anglos, males and females, were more critical of parking than were Negroes.

Fifty-nine percent of the Negro male respondents did not think that race relations were very good in Racine as contrasted to 14 percent of the Anglo males; but it must be emphasized that 87 percent of the Negro males and 70 percent of the Negro females said there was also discrimination in their hometown and whereas only 37 percent of the Negro males stated that race relations were good or very good in Racine, 58 percent of the Negro females responded in this fashion. The difference in responses by sex could be due to the greater number of interracial contacts of the Negro males as compared with females. Males also tended to be employed in the industrial sector of the economic institution whereas females are in commercial establishments.

Conclusion

The numerous individual questions and scales just discussed added up to one conclusion. Since leaving the Southwest or the South, the Mexican-Americans and Negroes had greatly improved their positions in terms of the social spaces that they occupied in the larger society. They perceived themselves as having moved upward as a consequence of their geographical mobility though they actually were far from the levels of living of the Anglos in the larger community.

The series of scales that has been presented in this chapter make it unmistakably clear that the life styles of Anglos and Negroes or Mexican-Americans was significantly different, although none revealed sharp differences between Mexican-Americans and Negroes. While it cannot be said that the way of life of the Mexican-American and Negro are entirely different, their relative poverty and deprivation forming a common bond, neither should it be concluded that life in the urban industrial milieu has resulted in reducing their disparate ways of coping with their social and physical environment to a common denominator dictated in its entirety by their position in the social system. Furthermore, the nature of the scales that were constructed failed to detect all of the differences between groups and within groups that would be observed by one who is trained to appreciate all aspects of cultural variation.

Mexican-Americans and Negroes were relatively satisfied with Racine in comparison to their hometowns and few had plans to return at some future date. In a variety of ways respondents indicated that the push from their hometowns had been sufficient to prepare them for some hardships during the process of adjustment in Racine. And while relatively pleased with their improved position in life, they had definite notions as to how the community could be changed in order to better satisfy the needs of persons who had been pushed from the Southwest and the South.

As a measure of cultural integration, we found that Mexican-American adults used Spanish predominantly when conversing with each other, while Mexican-American children spoke English almost entirely. Mexican-American respondents admitted that ethnic customs were declining but refused to be wholly critical of the behavior of those who had not yet acquired the middle class norms of the larger society.

Although the Anglos who had come to Racine had achieved far more economically than the Mexican-Americans and Negroes, from a standpoint of relative mobility and relative improvement, their way of life has probably not changed as much as had that of the Negro or Mexican-American.

NOTES

1. William H. Sewell, *The Construction and Standardization of a Scale for the Measurement of the Socio-Economic Status of Oklahoma Farm Families* Technical Bulletin No. 9, Oklahoma Agricultural and Mechanical College Agricultural Experiment Station, Stillwater, Oklahoma, April, 1940.

2. For example, see: William H. Sewell, "A Short Form of the Farm Family Socio-Economic Status Scale," Rural Sociology, Vol. 8, No. 2, June, 1943, pp. 161-170; John C. Belcher, "Evaluation and Restandardization of Sewell's Socio-Economic Status Scale," Rural Sociology, Vol. 16, No. 3, September, 1951, pp. 246-255; John C. Belcher and Emmit F. Sharp, *A Short Scale for Measuring Farm Family Level of Living: A Modification of Sewell's Socio-Economic Scale,* Technical Bulletin No. T-46, Oklahoma Agricultural Experiment Station, September, 1952; Emmit Sharp, "Criteria of Item Selection in Level of Living Scales," Rural Sociology, Vol. 28, No. 2, June, 1963, pp. 146-164.

3. The problem and a partial solution are presented in Charles S. Ramsey and Jenaro Collazo, "Some Problems in Cross-Cultural Measurement," Rural Sociology, Vol. 25, No. 1, March, 1960, pp. 91-106.

4. Although we do not have 1960 Census breakdowns for the availability of telephones in the homes of Mexican-Americans, Negroes, and Anglos in Racine and in communities of origin, it is of interest to note that 91 percent of the occupied households in Racine had telephones while only 39 percent in Union County and only 32 percent in LaSalle County had them. U.S. Census of Housing, HC(1) No. 51, Table 16; No. 45, Table 30; No. 26, Table 30.

5. Or, as Spilerman has stated, "Television must also be credited with stimulating the development of racial consciousness in Negroes. Sights of the insurrection of black persons elsewhere, or of Negroes being set upon by dogs, beaten, or worse, have enabled them to share common experiences, witness a common enemy, and in the process develop similar sensitivities and a community of interest." Seymour Spilerman, "The Causes of Racial Disturbances: A Comparison of Alternative Explanations," American Sociological Review, Vol. 35, No. 4, August, 1970, p. 646.

6. Ninety-five percent of the occupied households in Racine had a television set while only 82 percent in Union County and only 56 percent in LaSalle County had one. U.S. Census of Housing, HC(1) No. 51, Table 16; No. 45, Table 30; No. 26, Table 30.

7. Eighty-five percent of the occupied households in Racine had a washing machine, while only 81 percent in Union County and only 61 percent in LaSalle County had one. U.S. Census of Housing, HC(1) No. 51, Table 16; No. 45, Table 30; No. 26, Table 30.

8. An automobile was available to 81 percent of the residents of occupied households in Racine but 73 percent in Union County and only 55 percent in LaSalle County. U.S. Census of Housing, HC(1) No. 51, Table 16; No. 45, Table 30; No. 26, Table 30.

9. Grebler, Moore, and Guzman, *The Mexican-American People,* New York: The Free Press, 1970, pp. 254-258, show 62.5 percent of the Anglo-occupied dwelling units in the Southwest metropolitan areas were owner-occupied in comparison with 53.3 percent for Spanish surname persons and 42.9 percent for nonwhites. Between 1950 and 1960 Spanish surname home ownership increased more rapidly than the Anglo increase. Exactly what this means is difficult to say as there are some problems involved in the use of Census data. It may well be that Anglo home ownership had reached the leveling-off point and that Spanish surname ownership was simply at that point where rapid increase was still possible. Chapter 11 of this volume is devoted to a presentation and analysis of data on the generally poor housing conditions (overcrowding and substandard housing) of Spanish surname persons in the Southwest.

10. A few comparisons based on U.S. Census data may be helpful at this point. Ninety-eight percent of the housing units in Racine had hot and cold running water in the house but only 71 percent of the dwelling units in New Albany and 44 percent in Union County had these conveniences. Furthermore, only 37 percent of the households in Cotulla and LaSalle County were thus equipped. In Racine 95 percent of the households had a flush toilet while 72 percent of those in New Albany, 42 percent in Union County, and 43 percent in Cotulla and LaSalle County had what has for many years been considered more than simply a modern convenience. U.S. Census of Housing, HC(1) No. 51, Table 13; No. 45, Tables 25 and 28; No. 26, Tables 25 and 28.

11. For an excellent historical perspective on Chicago, see Allan H. Spear, *Black Chicago: The Making of a Negro Ghetto, 1890-1920,* Chicago: The University of Chicago Press, 1967.

12. The reader who is not familiar with the literature on the measurement of residential segregation should read Appendix A, pp. 195-245, Karl E. Taeuber and Alma F. Taeuber, *Negroes in Cities,* Chicago: Aldine, 1965.

13. Paul Hatt has shown this for certain areas of Seattle in "Spatial Patterns in a Polyethnic Area," American Sociological Review, Vol. X, No. 3, June, 1945, pp. 352-356.

14. Taeuber and Taeuber, op. cit., p. 34.

15. Taeuber and Taeuber, op. cit., p. 33. Milwaukee, Wisconsin had an index of 88.1, Chicago 92.6, Jackson, Mississippi 94.2, and San Antonio, Texas 90.1.

16. Grebler, Moore and Guzman, op. cit., pp. 271-289, contains an excellent discussion of the various bases for residential segregation of Mexican-Americans in the Southwest. One must conclude that the historical basis for segregation is not the same for old communities in the Southwest as for some newer metropolitan areas in the Southwest and in the industrial North.

17. *County and City Data Book, 1967,* U.S. Department of Commerce, Bureau of the Census, Table 2.

18. Some idea of the differences in housing conditions in general between Racine, New Albany, and Cotulla may be obtained by considering the differences in the proportion of deteriorating and dilapidated housing in each community. In Racine, 7.6 percent of the homes were deteriorating in 1960, 16.8 percent in Cotulla and 18.2 percent in LaSalle County and when deteriorating and dilapidated housing were added together, 9.0 percent of Racine's housing was in this category, as was 37.9 percent in Cotulla, 37.3 percent in LaSalle County, 39.3 percent in New Albany, and 46.9 percent in Union County. U.S. Census of Housing, HC(1) No. 51, Table 13; No. 45, Tables 25 and 28; No. 26, Tables 25 and 28.

19. While there are special problems involved in comparing estimated dwelling unit rates, one should at least note that the median value of owner occupied housing in Racine was $13,800, in New Albany $9,600 (Union County $7,400), and $5,000 in both Cotulla and LaSalle County. Median rentals were $81 for Racine, $45 for New Albany, and $38 for Cotulla. U.S. Census of Housing, HC(1) No. 51, Table 17; No. 45, Table 30; and No. 26, Table 30.

20. Perhaps the most startling contrasts between Racine and Union County or LaSalle County (data for New Albany and Cotulla being unavailable) were the two Census dwelling unit vacancy rates. Only 2.6 percent of Racine's dwelling units were vacant the year round but 6.9 percent of Union County's and 13.9 percent of LaSalle County's were vacant the year round. If vacancies were counted as of the time of the 1960 Census, then 2.8 percent were vacant in Racine, 6.3 percent in New Albany, 9.9 percent in Union County, 20.6 percent in Cotulla, and 21.4 percent in LaSalle County, an even starker contrast. Of these which were vacant in Racine on a year round basis, 2.6 percent were dilapidated. In Union County 35.7 percent of the year round vacancies were dilapidated and in LaSalle County the figure was 24.6 percent. U.S. Census of Housing, HC(1) No. 51, Table 12; No. 45, Table 28; No. 26, Table 28.

21. The fact that the Negro community in the South is so easily distinguished from the white community by the type of home construction, as well as other features, has been mentioned in the literature for several generations. Distinguishing between the Negro and the Mexican-American community in the Southwest, simply on a basis of housing, is not quite so easy. In reference to the Negro community see: Alphonso Pinkney, *Black Americans,* Englewood Cliffs: Prentice-Hall, Inc., 1969, Chapters III and IX. For a more general evaluation of housing in the United States, see: Nathan Glazer, Chapter 19, "Housing Problems and Housing Policies," *Metropolis in Crisis: Social and Political Perspectives,* Hadden, Masotti and Larson (eds.), Itasca, Illinois: Peacock Publishers, 1967, pp. 234-265. Although not pertinent to the Racine situation, Patricia Cayo Sexton's comments on public housing in Harlem indicate that a solution may not be a solution, i.e., the bulldozer and high-rise public housing destroy both slums and a valued social organization. Patricia Cayo Sexton, Chapter 4, "Urban Renewal: The Bulldozer and the Bulldozed," *Spanish Harlem: An Anatomy of Poverty,* New York: Harper & Row, 1965.

22. In reference to installment purchasing in 1959, it should also be remembered that not only were the Anglos of lower socioeconomic status than the 1960 Anglos, but the Anglos in 1959 were older than the Mexican-Americans in 1959 and this may play a part in the fact that they were not buying much on installment as compared to the Mexican-Americans.

23. In 1960, a scale was constructed by combining the three questions dealing with language. The respondents who demonstrated the least linguistic integration had the following characteristics: both husband and wife spoke Spanish, mostly Spanish or half-Spanish and half-English; people in Racine closest to the respondent spoke Spanish or mostly Spanish; children of the respondent spoke Spanish, mostly Spanish, or half-Spanish and half-English. Twenty percent of the respondents were at the English-speaking end of the scale and 33 percent were at the Spanish-speaking end of the scale; 19 percent were in the speaks mostly English category, and 29 percent in the speaks mostly Spanish category. Thus, the data revealed that the Mexican-Americans were skewed toward the Spanish-speaking end of the scale. This scale had a coefficient of reproducibility of .9181 and a minimum coefficient of reproducibility of .6300.

24. Thomas J. Scheff, "Changes in Public and Private Language Among Spanish-Speaking Migrants to an Industrial City," International Migration, Vol. 3, Nos. 1-2, 1965, pp. 78-84.

25. When Grebler, Moore and Guzman, op. cit., p. 384., asked San Antonio Mexican-Americans, "Is there anything about the Mexican way of life that you would particularly like to see your children follow?", 32 percent listed the Spanish language and 38 percent stated that they would like to see the retention of Mexican manners and customs. Also see pp. 423-432 of Chapter 18, the section on "The Persistence of Spanish." What Grebler et al., report for the Southwest gives us some idea of how the Mexican-American arrived in Racine bereft of English, particularly if from the lower income levels.

26. In 1959 numerous items on ethnicity were combined to form two scales. The first scale was designed to measure the ethnic practices and the second scale was designed to measure the ethnic orientation of the Mexican-American respondents. They were scalable in the Guttman sense and reproducibility was over 20 percent greater than that possible from the marginals. In a later section we shall discuss the relationship of ethnic practices and ethnic orientation to other variables. At this point, we wish only to emphasize that it was possible to distribute 1959 respondents on continua of this nature in a meaningful fashion.

27. Grebler, Moore and Guzman, op. cit., pp. 361-362, suggests that traditional patterns of decision-making are changing, particularly among the young, even in the southwestern states. When the statement, "A husband ought to have complete control over the family's income," was presented, 75 percent of those over 50 in San Antonio agreed but only 52 percent of those under 30 agreed.

28. The 1959 data on perception of changes was described in detail with breakdowns by sex, size of community of origin, occupation of father, etc., in Lyle W. Shannon, Chapter 11, "Adjustment of Rural Migrants in Urban Areas," Labor Mobility and Population in Agriculture, Ames: Iowa State University Press, 1961, pp. 122-149. External and internal measures of adjustment were compared in Lyle W. Shannon and Kathryn Lettau, "Measuring the Adjustment of Inmigrant Laborers," The Southwestern Social Science Quarterly, September, 1963, pp. 139-148.

29. The scale had a coefficient of reproducibility of .9470 and a minimum coefficient of reproducibility of .7620.

30. The scale had a coefficient of reproducibility of .9610 with a minimum

coefficient of reproducibility of .6980, indicating almost 30 percent improvement in predictability over the modal category of the marginals.

31. Negroes in metropolitan areas (New York City, Chicago, Atlanta, and Birmingham) were asked, "Do you think Negroes are better off in the South, in the North, or isn't there any difference?" Sixty percent of the metropolitan Negroes selected the North and only 8 percent the South. New York and Chicago Negroes, 57 and 55 percent respectively, selected the North. But only 32 percent of the Atlanta and 20 percent of the Birmingham Negroes selected the North. See Gary T. Marx, *Protest and Prejudice,* New York: Harper & Row, 1967.

32. Whether the differences were statistically significant from year to year also depended on the cutting point employed in testing for significance.

33. The Perception of Discrimination Scale has a coefficient of reproducibility of .9404 and a minimum coefficient of reproducibility of .8357. When dichotomized, the Negro/Anglo $\chi^2 = 42.98$, 1 d.f., $p < .001$.

34. Negro efforts to secure better housing have been so dramatized by the media of mass communication in recent years that the reader may wonder what contribution is made by even mentioning this typical response. The question here is just a bit different in that it is directed toward the Negro inmigrant whom we have stated has probably bettered himself in housing by a move to the North. As in all of the responses to questions about change or questions requiring an evaluation of existing institutions in Racine, critical responses may be interpreted as evidence of cultural integration—as in increasing awareness of one's position in reference to those of persons in the larger society. On the other hand, it would be naive to believe that the southern Negro has aspirations which are only slightly above his present level. Although the housing aspirations of the Negro are below those of the white, they are considerably higher than the housing he has, even among those who still reside in the South. See John C. Belcher, "Differential Aspirations for Housing Between Blacks and Whites in Rural Georgia," Phylon, Vol. XXXI, No. 3, Fall, 1970, pp. 231-243. Aspirations are only part of the picture in housing as well as jobs. For a general picture of the struggle for civil rights and for housing, see Part V, *The Negro in Twentieth Century America,* John Hope Franklin and Isodore Starr (eds.), New York: Random House, 1967, or Charles Abrams, "The Housing Problem and the Negro," and Eunice and George Grier, "Equality and Beyond: Housing Segregation in the Great Society," in *The Negro American,* Talcott Parsons and Kenneth B. Clark (eds.), Boston: Beacon Press, 1965, pp. 512-554.

35. Job opportunities were emphasized by Negroes and Mexican-Americans in Racine at every opportunity presented in the interview schedule. Lack of job opportunities was their reason for leaving the former place of residence, their reason for coming to Racine, their reason for being more satisfied in Racine than in former place of residence, the thing that they would like to see more of in Racine, and the area in which they felt Racine to be most deficient. That this is not a new finding is evidenced by the fact that almost every volume on the Negro or the Mexican-American not only dwells on their employment status by occupational distribution and income but also on their awareness and perception of the situation. See, for example: Pinkney, op. cit., Chapter IV.

The National Advisory Commission on Civil Disorders sponsored attitude surveys in 20 cities in 1967. In the 1,200 interviews conducted, unemployment and underemployment were found in 19 cities, placing employment at the top of the list, followed closely by police practices, inadequate housing, inadequate education, and poor recreational facilities. See *Report of the National Advisory Committee on Civil Disorders,* New York: Bantam Books, 1968. Also see Angus Campbell and Howard Schuman, *Racial Attitudes in Fifteen American Cities,* Survey Research Center, Institute for Social Research, The University of Michigan, June, 1968. This is a preliminary report on a survey of more than 5,000 Negroes and whites in 15 cities for the National Advisory Committee on Civil Disorders. Seventy-two percent of the males "believe many or some Negroes miss out on jobs today because of discrimination." and 68 percent "believe that many or some Negroes miss out on promotions today because of race." In respect to themselves, 34 percent reported having been refused a job because of racial discrimination and 18 percent reported having been refused a promotion because of racial discrimination. In respect to the question of housing, 75 percent stated that "many" or "some" (as against "few") Negroes in this city miss out on good housing because of racial discrimination. Fewer whites than Negroes perceived those patterns of discrimination against Negroes.

On the other hand, the idea of shared disagreements by most Negro Americans as the basis for demonstrations and riots has been rejected by Seymour Spilerman in "The Causes of Racial Disturbances: A Comparison of Alternative Explanations," American Sociological Review, Vol. 35, No. 4, August, 1970, pp. 627-649. But in turn, he concludes that the larger the Negro population the greater the likelihood of disorder.

36. Although the three-point scale (generally satisfied, somewhat satisfied, very dissatisfied) utilized by Campbell and Schuman, op. cit., for evaluating parks and playgrounds for children was not identical to our response categories, the results were somewhat the same for their sample of more than 5,000 Negroes and whites in 15 U.S. cities. The Anglo males and females were skewed toward a more favorable evaluation of parks and playgrounds in both studies than were the Negroes.

37. Although the interviews turned up no particular concern about shopping facilities for Negroes, it is possible that the situation in Racine differs from other urban areas where members of minority groups are more dependent upon "ghetto" or "poverty" area stores. It appears that both ghetto store pricing and minority group status contribute to higher retail prices for Mexican-Americans and Negroes. For a Los Angeles example, see Frederick D. Sturdivant and Walter T. Wilhelm, "Poverty, Minorities, and Consumer Exploitation," Social Science Quarterly, Vol. 50, No. 4, March, 1970, pp. 1064-1071, and for the earlier New York data see David Caplovitz, *The Poor Pay More,* New York: The Free Press, 1963.

38. Public schools were evaluated by the Campbell-Schuman samples, op. cit., utilizing the same three-point scales as that for parks and playgrounds. Anglo males were more satisfied than were Negro males, but the females differed very little, as in the Racine case.

Chapter 8

CHILDREN AND CHILD-REARING PRACTICES

Introduction

In the last chapter we touched upon several aspects of the cultural integration of inmigrants, such as changing language usage and the general decline of "ethnic practices." Closely related, for that matter intertwined, is the topic of children and child-rearing practices, a most convenient substantive area in which to examine intergenerational cultural integration. Although our basic interest is in the process of socialization as it relates to intergenerational change, we are also concerned about adult perceptions of the role of the family and other institutions in the socialization of children, adult impressions of occurrences in their own youth as compared to what was happening to their children in 1960, and with adult conceptions of what "should have happened" as opposed to the intergenerational change that actually occurred.[1]

In the course of the interview parents were probed about the importance of their role in both formal and informal settings for socialization of the child. What did they know of the experiences of their children in and outside the school? What educational and occupational aspirations did they have for their children and what were they doing to help them reach these goals? Perhaps more important, how did they evaluate the chances of their children of reaching these verbalized goals or how satisfied would they be if these goals were not reached?

Intergenerational Differences

In 1959, respondents were asked three questions about children: (1) "Are children more or less obedient nowadays?" (2) "Are children more or less respectful nowadays?" and (3) "Do you think children will be as willing to work as hard as their parents did?" Both Anglo and Mexican-American respondents described today's children as being less obedient and less respectful, but 78 percent of the Mexican-Americans as compared to 62 percent of the Anglos, a significant difference, described them as less willing to work as hard as their parents had as children.

Zula Patino, mother of six children, had definite ideas about how her children would respond to work:

> "If they had a good head they naturally would like to work but the majority don't like to work."

The answers of Martin Lopez indicated that members of our sample were now turning away to a certain extent from the traditional Mexican-American concept of hard physical labor as a natural and noble way of earning a living. He seemed to comprehend that changing times and places alter man's traditional patterns of work.

> "In our generation we had to do field work. Now they get jobs in a store or something easier."

> "No, they all will have better jobs and education. Hard labor is being done away with and when they're old enough to work they don't have to work too hard."

When respondents were asked how their lives differed from those of their parents both Mexican-American and Anglo females pictured parental discipline as stricter more often than did the males. Approximately 10 percent mentioned greater educational opportunities for themselves, slightly more than 10 percent mentioned better occupational opportunities, and less than 10 percent mentioned greater material possessions in response to specific probes. In general, neither Mexican-Americans nor Anglos perceived of themselves as having changed as much from their parents as their children had from them.

The Socialization Process

In earlier chapters we hypothesized "hometown" geographical area of socialization to be an important determinant of the degree to which the Mexican-American, Negro, or Anglo adjusted in the northern industrial community. More definitively, we hypothesized that socialization in a *rural* southwestern or southern community constituted a handicap to economic absorption and cultural integration in addition to the handicap of being a Mexican-American or Negro. If the process of socialization in a given subculture resulted in the acquisition of a distinctive set of beliefs, values, goals, and approved means of implementing them, then each of our samples could be expected to respond differently to a series of questions that probed the process and content of socialization.

Formal Versus Informal Education

The Role of the Home

A series of questions about the function of formal versus informal education was asked each year of the survey. In 1960, respondents were asked the following question: "One sometimes hears the saying, 'education begins in the home.' In your opinion, what are the most important things a boy must be taught in the home?" Seventy-one percent of the Mexican-Americans and 63 percent of the Anglos most frequently mentioned the teaching of respect or obedience. Significantly more Mexican-Americans (57 percent) than Anglos (45 percent) mentioned the need to teach social norms.[2] Anglos (32 percent) were more likely to mention the teaching of religion than were Mexican-Americans (19 percent), a significant finding. Mexican-Americans (36 percent) were also more likely than Anglos (19 percent) to cite the home as the place for teaching educational and work aspirations.

Two Guttman scales of home functions (Table 34) were constructed, one for males and one for females. A respondent who believed in maximum emphasis on the home in the process of socializing the male would have enumerated the teaching of respect and obedience, the acquisition of social norms, the acquisition of educational and work aspirations, and some other characteristic such as religion. The difference between Mexican-Americans and Anglos was statistically significant; although both were fairly evenly distributed across the continuum of scale scores, the Anglos were skewed toward the lower end, placing less emphasis on the home as a place of socialization.

TABLE 34. PARENTAL PERCEPTION OF CHILD-REARING FUNCTION OF THE HOME, 1959 (IN PERCENTAGES)

		Male Children		Female Children	
Scale Type		Mexican-American	Anglo	Mexican-American	Anglo
Minimum emphasis on home	0	7	16	19	23
	1	25	30	56	26
	2	31	31	19	25
	3	25	12	4	20
Maximum emphasis on home	4	12	11	2	6
		100	100	100	100

Male Children: $R = .8870$; $MR = .6830$; Mexican-American/Anglo $\chi^2 = 8.77$, 1 d.f., $p < .01$.

Female Children: $R = .9180$; $MR = .7600$; Mexican-American/Anglo $\chi^2 = 28.43$, 1 d.f., $p < .001$.

Manuel Fuentes, father of three boys, ages 10, 11, and 13, felt that a boy should not be allowed to do whatever he pleased in the home:

"The saying says that a child is like a tree one plants. If it isn't properly cared for it will grow wrong."

Ramond Segura, father of three girls and one boy aged 8, felt that

"The home should provide a boy with good 'paternal' education such as to respect and not to steal."

Another question asked of the Mexican-Americans in 1959, "In what ways does a father teach his son to be a good man?" elicited such replies as: "set an example," "give advice," "teach." In only 2 percent of the cases did respondents mention negative techniques of control, such as punishment, prevention, or correction.[3]

When respondents were asked what a *girl* must be taught in the home, housework skills was most frequently mentioned (Mexican-American 73 percent and Anglos 59 percent) with respect or obedience and social norms following (Mexican-Americans 55 percent and Anglos 50 percent for each). Anglos were next most likely to mention the need to teach religion (29 percent) in contrast to Mexican-Americans (4 percent). Neither Mexican-Americans nor Anglos were apt to mention educational or work aspirations for girls.

Turning again to Table 34, persons placing maximum emphasis on the home as a socializing agency for females mentioned the teaching of household skills, the teaching of respect and obedience, the acquisition of educational and work aspirations, and other functions such as the teaching of social norms and religion. The Mexican-Americans placed somewhat less emphasis on the home in socializing females than did the Anglos. In sum, Mexican-American respondents placed less emphasis on the home as a place of socialization for the child than would have been expected. Although the difference between the two ethnic distributions was statistically significant, it was in one direction for males and the other for females. Why did the Mexican-Americans place less emphasis on the function of the home in training females than Anglos? Why was there practically no difference in the responses of males and females (the Mexican-American in particular) to the questions on which these scales were based?[4] Although none of the respondents expressed greater educational ambitions for males than for females nor indicated in any way that hopes and aspirations for their children were really centered on the males, we sometimes received this impression from the Mexican-Americans. Nevertheless, we are inclined to agree with those who say that many social scientists are full of misinformation about Mexican-Americans.[5] What they believe to be going on in the heads of Mexican-Americans has been influenced by what is going on in their own heads.

The Role of the School

In 1960, respondents were asked the following questions: "It is often said that school is only one of the places that children learn. In your opinion, where do children learn more—in school, in the home, or when they go to work?" "In the long run, which training is more valuable for the child?" The data indicated a seemingly unexpected response pattern. Mexican-Americans (74 percent) placed the school significantly higher than either Negroes or Anglos while Negroes (46 percent) placed the school significantly higher than the Anglos (29 percent). On the other hand, 62 percent of the Anglos placed the home higher than either Mexican-Americans or Negroes and 48 percent of the Negroes placed the home higher than Mexican-Americans (22 percent), each significantly different. Jobs were seldom set forth as a place for training the child, but were mentioned by 21 percent of the Anglos, 10 percent of the Negroes, and 6 percent of the Mexican-Americans.

That Mexican-Americans placed the greatest emphasis on the school and the least on the home reflected, it is believed, an increasing hope among minority groups that education would be the key to occupational mobility. Mexican-American and Negro respondents must not have been cognizant of or at least been undismayed by the ceiling on occupations that so often exists for them in the industrial community regardless of their educational qualifications. For many Mexican-Americans and Negroes it may be a case of not yet having reached that level of educational attainment at which they would experience occupational restrictions. Another study conducted at approximately the same time revealed that Negroes desired education in order to enter the professions where they would not be so vulnerable to discrimination as in other situses or occupational hierarchies.[6] Miller and Riessman have proposed that members of the working class strongly desire education for their children since they recognize their lack of education as an outstanding weakness.[7] In any event, the Racine respondents were quite obviously in the same category, and presented the view of persons in the working class subculture who are mobility oriented regardless of race or ethnicity.

Attitudes Toward Formal Education

Education and Sex

In order to test the hypothesis that the educational aspirations of Mexican-Americans vary with the sex of the child, in 1959 respondents were asked if girls should have as much schooling as boys. Contrary to expectations, 58 percent of the Mexican-Americans answered in the affirmative, an answer that was slightly less but not significantly different from the 65 percent of the Anglos who responded in this fashion. Anglos gave a predominantly economic rationale for their position, whereas the Mexican-Americans gave more general answers, such as:

". . . A girl should receive as much schooling as a boy because it is going to be a woman's world . . . because this nation employs just as much women as men. If you notice the offices, there are more women than men."

". . . I believe so because schooling is the most important thing."

". . . Yes, so the nation can have an equal number of smart people and just as good a worker in both sexes."

". . . The girl should try to go to college because it would help them teach their children better living conditions."

In 1960 respondents were asked if they would like to have more education for some children than for others. Twenty-five percent of the Anglos, 10 percent of the Mexican-Americans, and 10 percent of the Negroes preferred more for boys. Twelve percent of the Anglos, 5 percent of the Mexican-Americans, and 4 percent of the Negroes specified more for the bright ones. Similarly, 12 percent of the Anglos but only 2 percent of the Mexican-Americans and 4 percent of the Negroes indicated that it depended on the child's aspiration.

The Function of the School

When, in 1960, respondents were asked to compare the importance of education in their day with the present, 90 percent of the Negroes, 88 percent of the Anglos, and 79 percent of the Mexican-Americans considered it more important in 1960. When asked why they thought education was more important today, 74 percent of the Negroes, 61 percent of the Anglos, and 54 percent of the Mexican-Americans referred to increased competition for jobs or ability to get a better job.

The 1960 data were used to construct a scale summarizing attitudes toward the function of schools. The respondent who scored highest on the scale was the person who attributed the most importance to education; he believed that education was more important than when he or she was in school, that the schools should prepare people for an occupation, that the most valuable training for a child was in school, and he mentioned occupation twice in responding to the question on function of the schools. The data in Table 35 indicate that the Mexican-Americans (68 percent were in the two highest scale categories) differed significantly from the Anglos and Negroes in placing the most positive evaluation on the role of education. Negroes in turn (44 percent) scored significantly higher than the Anglos.

In an attempt to discern what education meant to Mexican-Americans, Anglos, and Negroes, they were asked in 1960: "Some parents feel that many of the things that are taught in school are a waste of time. What do you think are the main things that the school should teach?" Almost 90 percent had no complaints about what was being taught in the schools. Of those who responded positively to the questions, 62 percent of the Anglos emphasized the need for a basic academic program in contrast to 56 percent of the Negroes and 40 percent of the Mexican-Americans. Sixteen percent of the Mexican-

TABLE 35. PARENTAL ATTITUDE TOWARD FUNCTION OF SCHOOLS, 1960 (IN PERCENTAGES)

Scale Type		Mexican-American	Negro	Anglo
Education least important	0	12	4	4
	1	10	27	24
	2	10	25	44
	3	58	39	24
Education most important	4	10	5	5
		100	100	100

R = .9069; MR = .6925; Mexican-American/Anglo χ^2 = 40.90, 1 d.f., p $<$.001; Negro/Anglo χ^2 = 9.20, 1 d.f., p $<$.01; Mexican-American/Negro χ^2 = 27.87, 1 d.f., p $<$.001.

Americans, 10 percent of the Negroes, and 7 percent of the Anglos emphasized social adjustment and discipline as school functions.[8] Because of the possibility of variation in the intensity of responses among the samples, interviewers were requested to characterize responses according to the degree to which they represented well-formulated beliefs. Sixty-eight percent of the Anglo sample was judged to have well-formulated beliefs but only 54 percent of the Negroes and 53 percent of the Mexican-Americans.

In 1961 almost 90 percent of the respondents agreed that Racine schools were doing a good job. When asked whether there was anything which they would like changed in the schools, the most frequently mentioned category of concern by the Negroes (10 percent) was the relative absence of Negro teachers, whereas 14 percent of the Anglos were more interested in increasing the academic emphasis.

Involvement of the Parents

Our next effort was to ascertain specific indications of involvement of the respondents in the educational experiences of their children. In 1960, when asked whether any of their children ever needed help with their homework, 69 percent of the Negroes, 64 percent of the Anglos, and 51 percent of the Mexican-Americans answered in the affirmative. The 1961 data, consistent with the 1960 findings, indicated no significant difference between Anglos (62 percent) and Negroes (68 percent). When respondents were asked who helped children with their homework in 1960, 40 percent of the Anglos, 37 percent of the Negroes, and only 24 percent of the

Mexican-Americans, and in 1961, 51 percent of the Anglos and 37 percent of the Negroes mentioned parents (either singly or in combination). The comparatively low percentage of Mexican-American parents who help their children with their homework might seem to indicate a lack of interest in being involved in their children's education but a more applicable conclusion would be that the inadequacy of their parents' educational background precluded assistance—under such circumstances children would neither ask their parents for help nor would parents feel themselves capable of helping. While fewer of the Mexican-American children (42 percent) received good grades than either the Negro (49 percent) or the Anglo 52 percent), as reported by the parents, the difference was not statistically significant.

Social Participation of Children as Reported by Parents

In 1961 we asked Anglo and Negro respondents about the leisure time activities of their children. Our knowledge of the family structure of the Negro families led us to expect that the females would be more aware of the organizational membership of their children than the males.[9] As it turned out, there was not much difference between male and female responses in either the Negro or the Anglo sample but there was a significant difference in the pattern of membership. Anglo children (20 percent) were more likely than Negro (11 percent) to belong to church groups, or to Scouts (Anglos 21 percent and Negroes 13 percent), while Negro children were more likely to belong to the YMCA or the YWCA (Negro 27 percent and Anglo 20 percent).

Respondents were also asked about after-school or school-related activities of their children. The most noticeable difference was the greater participation of Negroes (44 percent) in sports as contrasted to Anglos (25 percent). No activities were reported for 40 percent of each group, however.

Permissiveness of Parents

When asked where their children did their homework, half the Anglo parents in 1961 reported "in school" and half reported "at home" while two-thirds of the Negro parents reported "in school." Such data lead to a multitude of interpretations, one as reasonable as the next—the Negro home is not the place to do schoolwork, some

Negro parents do not pay enough attention to what their children are doing, Negro children have a different conception of whether or not they have homework to do, or Negro children are presenting their parents with the impression that they are completing the homework in school, whether they are or not.[10]

The common presumption that Negroes are more permissive than Anglos in parental control of children's activities seemed to be fallacious, for over 90 percent of either Negro or Anglo respondents required their children to be in at certain times on week nights and weekends.[11] Almost all the Anglo and Negro parents believed that they ought to know what their children were doing with their spare time and although there were a variety of reasons why they thought they should know, no real pattern of differences emerged.

The responses to the question of how old children should be before parents no longer need be concerned about their leisure time were scattered; one-fifth of the Anglos and well over one-third of the Negroes were of the opinion that children ought to be at least 18 to 20 years of age and another fifth of the Anglos felt that parents ought to keep close tab on their children until they have left home regardless of age.

More of the Anglo children (25 percent) were reported to have part-time jobs during the year than were Negro (12 percent), which could be interpreted merely as an indication of awareness of parents of the employment of their children or of the general dearth of jobs for Negro children, rather than the inclination of Negro children to seek part-time jobs. Over 40 percent of the Negro children attended movies as often or more often than once every two weeks while only 24 percent of the Anglo children fell into this category. Negroes also viewed TV significantly more often than did Anglos.

Interviewers concluded this section of the interview by judging whether the respondents had a definite knowledge of their children's activities. Almost no difference was evident between Anglos, Negroes, and Mexican-Americans, but 87 percent of the female respondents were categorized as having definite knowledge of their children's activities as compared to 78 percent of the males.

Level of Educational and Occupational Aspirations

Among the social psychological variables that are hypothesized to intervene between antecedent sociological variables and integrative behavior is the level of aspiration that adults have for themselves and their children. We hypothesized that level of aspiration is related to the antecedent sociological variables but not entirely determined by background and early life experiences.[12] More recent events in the lives of inmigrant minority group members could raise or lower their aspirations markedly. For example, we learned that the project had left its mark on the community in terms of a rise in the level of aspiration of many who were interviewed and that the Mexican-American interviewers themselves had revised their goals upward as a consequence of participation.

There is an abundant literature on the value system of lower socioeconomic status groups, the lower class or the working class, the origin of the greater proportion of the Mexican-American and Negro samples. As stated previously, the working class subculture, although this is not a distinguishing feature, is held to place great value on education. Nonetheless, the educational and occupational aspirations of such parents for their children requires interpretation as it applies to our respondents. At the same time that we are told that the working class subculture values education, it is also indicated that the lower the "social class," the lower the level of educational and occupational aspirations.[13] Using reference group theory to explain the differences in aspirations between members of various socioeconomic status groups, we would expect individual lower socioeconomic status groups to take the standards of their group as a reference point against which to develop their educational and occupational aspirations.[14] We must not expect what they say either about education or occupation to be put in the same terms as would a mobility oriented middle socioeconomic status category respondent. Thus far we have observed that general questions about education and schooling elicit favorable responses. The interviews made it clear that members of each group wished to better themselves and wished to see their children better themselves, but what was said about education and occupation need be interpreted within a framework of relative mobility.

Since the main goal of the lower class or working class tends to be the achievement of economic security, its members give such reasons as "pays well" or "good hours" for choice of occupation whereas congeniality of atmosphere and intellectual challenge are more likely to be advanced by members of the middle and upper classes.

It should not be surprising that educational and occupational aspirations for children were not always readily elicited from the inmigrants and that their conceptualization was not always as specific as hoped for in responses to the questions. While there appeared to be an almost universal desire to have a higher level of living and a willingness to leave former homes in search of it, many did not seem to envisage how much education was needful for their children or what type of work they preferred for them. Some Mexican-American and Negro and some Anglo respondents were only partially integrated into the complex, urban industrial milieu; because of limited education and urban experience they seemed to know little about the nature and organization of the society of which they were a part and were frustrated when they attempted to translate their vague aspirations and those for their children into concrete plans for mobility. Perhaps the execution of a move to a northern industrial community provided sufficient upward mobility to satisfy the aspirations temporarily of those inmigrants who gave generally satisfied responses.

In an attempt to discover respondents' expectations for their children we asked in 1959, "What about your children, in what important ways do you think they will be different from you?" Mexican-Americans and Anglos agreed that social and economic opportunities would be better and that their children's level of living would be higher, but Mexican-Americans (50 percent) differed from the Anglos (34 percent) to a significant extent in their predictions of greater educational opportunities for their children. Mexican-American males (60 percent) and Anglo (40 percent) were more likely than females in either category to arrive at this conclusion.

> "I'm hoping that they will have more school than I. I won't want them to work as hard as I have worked." (Mexican-American Male)

> "They will go more to school and learn the American way more than I did." (Mexican-American Male)

> "To learn more than us, they have everything that is necessary and it is easy to learn more modern." (Mexican-American Male)

> "They have more inspiration to do things and will probably have more wealth and knowledge than I." (Mexican-American Male)

> "Yes, they are different. In the first place they bring their sweethearts home—and I could never do that in my time. They go bathing, play cards, and everything." (Mexican-American Female)

Educational Aspirations

Should the reader wonder why the relationship of education to occupation is a constantly recurring theme in this volume, even though we have stated that the relationships found for Anglos are not obtained in the same degree for Mexican-Americans and Negroes, the relationship of socioeconomic status of parents to children's probability of obtaining higher education is so great that it must be carefully considered at this point in a somewhat different context than previously. No one has stated it better than have Sewell and Shah in "Socioeconomic Status, Intelligence, and the Attainment of Higher Education."

"The educational system plays an important role in the allocation of personnel to various occupational positions. It sorts people according to differences in valued abilities, channels them into streams of training which develop their capacities, and encourages them to aspire to adult roles that are in keeping with their talents. However, many factors other than the ability of the student influence his eventual educational experiences and attainments. These include differences in the level and quality of education available in the country, region, or community in which he lives; differential access to educational facilities according to his social class status, religion, race, and ethnic origins; differences in his motivations, values and attitudes; and differences in the willingness and ability of his parents and significant others to provide the financial and psychological supports necessary for the maximization of his talent potentials."

Sewell and Shah go on after a careful evaluation of data on Wisconsin high school graduates to state:

"Perhaps a more critical factor in the process of obtaining higher education is the decision to plan on and to enter college. At this point, over a fourth of the high ability males (those in the top quarter of the intelligence distribution) and almost half of the high ability females drop out of the process by not planning on or not entering college. Socioeconomic origins powerfully affect these decisions of high ability youth of both sexes; just over half (52.4%) of the high ability males of low socioeconomic status enroll in college in comparison with 90.7% of the high status males of equal ability; for females the corresponding percentages are 27.5 and 76.4%. Moreover, the yield of college graduates from high ability males is only 20.1% for those with low socioeconomic status origins in comparison with 64.0% from those with high socioeconomic status backgrounds; for females the yields of college graduates are 13.8 and 51.1% respectively. Even if only those who enter college are considered, socioeconomic status exerts a powerful influence; only 38.5% of the high ability males who are low in socioeconomic status graduate in comparison with 70.6% of those of equal ability but high in socioeconomic status. For females the respective figures are 50.0 and 66.9%. Similar trends hold for less able youth."

And as Sewell and Shah conclude,

"From all of this evidence it seems clear that although intelligence plays an important role in determining which students will be selected for higher education, socioeconomic status never ceases to be an important factor in determining who shall be eliminated from the contest for higher education in this cohort of Wisconsin youth."[15]

In each year of the survey respondents were asked how much schooling they desired for their children. The responses indicated educational aspirations which were often quite vague and unrealistic. In 1959, responses to the question, "How much schooling would you like your children to have?" were neither followed with sufficient probing nor coded so as to allow us to say other than that half of both the Mexican-Americans and Anglos wanted their children to have as much education as they could get or at least a college education. In 1960, the question on educational aspirations was preceded by other questions on education which apparently set the stage for more definitive responses. Therefore, as seen in Table 36, 25 percent of the Mexican-Americans aspired to a college education for their children in contrast to 50 percent of the Negroes and 67 percent of the Anglos.[16] But close to 30 percent of the Anglos still specified no level for their children, only that they receive as much education as possible.[17] These respondents were probably education-oriented but not sufficiently cognizant of what response would be appropriate. In 1961, Anglo respondents had slightly but not significantly higher educational aspirations for their children than did Negro respondents. The marked decline of non-specific

TABLE 36. EDUCATIONAL ASPIRATIONS FOR CHILDREN, 1960 AND 1961 (IN PERCENTAGES)

	1960			1961	
	Mexican-American	Negro	Anglo	Negro	Anglo
Grade school; junior high school	2	0	1	0	0
Complete high school	33	16	14	28	20
College and/or beyond	25	50	67	67	74
As much as he can get	29	30	11	3	3
As much as he wants; let the children decide	5	2	4	0	3
Not reported	6	2	2	2	1
	100	100	99	100	101

1960: Mexican-American/Anglo χ^2 = 27.98, 1 d.f., p < .001; Negro/Anglo χ^2 = .09, 1 d.f., not significant; Mexican-American/Negro χ^2 = 23.81, 1 d.f., p < .001.
1961: Negro/Anglo χ^2 = 2.83, 1 d.f., not significant.

answers might have been the result of a discussion of education that took place in the home or community following the 1960 interview which prepared those respondents who were reinterviewed in 1961. The 1961 responses represented, as it were, a certain amount of sharpening of opinion on the part of the respondents.

Follow-up questions were designed to ascertain how well the respondents had "thought out" their educational aspirations. In 1959, when asked "Why is that?" in reference to the aspirational responses which they had given, Mexican-Americans mentioned economic advantages to a significantly greater extent than any other single factor. Moreover, when other factors were mentioned they were combined with statements about economic advantages in over two-thirds of such instances.

Each question that has been discussed indicates that each group of Racine respondents was committed quite clearly to the increasing value of education; 80 to 90 percent of the respondents thought that education was more important today (circa 1960) than in their youth. At least half of each sample gave economic reasons. [18]

Keeping in mind this repeated commitment to the values of education and the linking of educational aspirations with economic goals, one cannot help but be surprised at the 1960 responses to an additional question, "It's sometimes hard to tell how things will actually work out. If things turned out that your children completed junior high school (9th grade), and then went to work, would you be satisfied, or dissatisfied?" The level of satisfaction for Mexican-Americans was the lowest (Table 37) and that for the Anglos was the

TABLE 37. LEVEL OF SATISFACTION FOR EDUCATIONAL ATTAINMENT OF CHILDREN, 1960* (IN PERCENTAGES)

	Mexican-American	Negro	Anglo
Junior high school	37	20	6
High school	42	39	47
Two years of college	9	10	14
College degree or higher	10	26	31
Not ascertained	3	5	2
	101	100	100

Mexican-American/Anglo χ^2 = 42.39, 1 d.f., p $<$ .001; Negro/Anglo χ^2 = 3.31, 1 d.f., not significant; Mexican-American/Negro χ^2 = 22.40, 1 d.f., p $<$ .001.

*The following question was asked four times with successively higher levels of education until respondents stated that they would be satisfied. "It's sometimes hard to tell how things will actually work out. If things turned out that your children completed junior high school (9th grade), and then went to work, would you be satisfied or dissatisfied?"

highest; that of Anglos and Negroes was not significantly different.[19] The necessity of obtaining a specified level of education for occupational reasons was given in more than half the cases by Mexican-Americans, Negroes, and Anglos as the reason for being dissatisfied if their children did not complete a stated level of schooling.[20] Education as an end in itself was seldom mentioned.

In order to approach the question of education still another way, to inject a note of realism by raising the question of money, the following question was asked: "You can't always tell about the way things will work out. Here are some statements. Tell me, as far as you can see, which statements would come the closest to the one that you would agree with. For a person in my financial position, it will be practically impossible to keep my children in school past the 9th grade, 12th grade, to put my children through college." Anglos were generally more optimistic than either Mexican-Americans or Negroes while Negroes were more optimistic than Mexican-Americans about keeping children in school, regardless of the level. Agreement was expressed by 17 percent of the Mexican-Americans, 12 percent of the Negroes, and 5 percent of the Anglos; only 8 percent of the Mexican-Americans, 4 percent of the Negroes, and none of the Anglos was undecided. When the question applied to the 12th grade, response patterns shifted with agreement shown by 27 percent of the Mexican-Americans, 14 percent of the Negroes, and 13 percent of the Anglos. The number of "undecided" answers increased to 15 percent for the Mexican-Americans or Negroes and to 4 percent for Anglos. The final question in reference to college, "For a person in my financial position, it will be practically impossible to put my children through college," was accepted by 17 percent of the Mexican-Americans, 18 percent of the Negroes, and 11 percent of the Anglos.[21]

The intriguing part of this entire series of questions is not the percentage who agree that they cannot keep their children in school beyond a certain level, but the proportion who disagree and systematically disagree in decreasing numbers as they reach progressively higher levels of education, as shown in Table 38. While the proportion who accepted the possibility of keeping their child in school was decreasing, the number who could not respond to the question and who were undecided was also systematically increasing. The significant differences that were found in separate tables for each question were based on the facts that (1) Mexican-Americans, Negroes, and Anglos progressively and in that order disagreed that

**TABLE 38. RESPONDENTS DISAGREEING THAT THEY COULD NOT KEEP CHIL-
DREN IN SCHOOL AT OR BEYOND STATED LEVEL, 1960 (IN PERCENTAGES)**

	Mexican-American	Negro	Anglo
Beyond 9th grade	64	81	86
Beyond high school	30	55	70
Send to college	12	20	42

they would be unable to keep their children in school beyond a given level, and (2) there was a systematic increase in those who could not respond to the question.

In 1961, respondents were asked how much education they thought their children would actually receive. While 52 percent of the Anglos responded, "a few years of college," or "college and beyond," only 20 percent of the Negro responses fell into this category; the difference between the two groups was statistically significant. In a follow-up question the respondents were asked whether they would be satisfied if their children finished a level of education one step lower than that which they expected. Over half the respondents, 62 percent of the Anglos and 55 percent of the Negroes, indicated that they would not be satisfied but the difference between the two groups was not statistically significant.

Occupational Aspirations

In each year of the survey respondents were asked what kind of work they would like their children to go into. In 1959, Anglos (22 percent) were more likely than Mexican-Americans (4 percent) to respond that the choice of an occupation should be left up to the children themselves. Although Mexican-Americans mentioned such positions as doctor, lawyer, engineer more frequently than did Anglos, they also, particularly Mexican-American females (29 percent), were more likely to mention unspecified white-collar work, "easy work," or "office work."

In 1960 (Table 39), the proportion of Anglo respondents who preferred to let their children decide their occupation was almost twice as large as that for Mexican-Americans or Negroes. Professional or similar prestige positions were chosen by 30 percent of the Negro parents, 25 percent of the Mexican-Americans, and 23 percent of the Anglos.[22] Nineteen percent of the Mexican-Americans (26 percent of the females and 11 percent of the males), mentioned white-collar

TABLE 39. OCCUPATIONAL ASPIRATIONS FOR CHILDREN, 1960 AND 1961 (IN PERCENTAGES)

	1960			1961	
	Mexican-American	Negro	Anglo	Negro	Anglo
Doctor, nurse	7	15	7	17	7
Priest, minister, nun	1	0	3	0	5
Teacher	3	5	2	1	4
Engineer	4	1	4	2	3
Entertainment (sports, music, acting)	1	4	1	2	1
Other professional (managerial, proprietor, technician)	9	5	6	7	4
White collar, unspecified, clerical, any easy job	19	4	2	4	4
Other job	8	4	4	3	4
Leave it up to the children, or children's abilities	26	22	46	17	41
Not reported	1	2	1	3	2
Inapplicable, no young children	21	39	23	45	24
	100	101	99	101	99

jobs in contrast to 2 percent of the Anglos and 4 percent of the Negroes. The remarks of Hyman and others who contend that a "let them decide for themselves" response usually means "let them decide for themselves as long as they select a profession or some other socially acceptable level as defined by 'middle class' persons" are cogent at this point. [23]

That 80 percent of the Anglos as compared to 63 percent of the Negroes and Mexican-Americans had well-formulated responses in the judgment of the interviewers must also be taken into consideration. Again, among those children "allowed" to make their own job decision (46 percent Anglo, 26 percent Mexican-American, 22 percent Negro) there is more likelihood of a choice of higher occupational levels because of background or congruous advantages. [24]

Responses to questions concerning level of aspiration for children may be in part indicative of the aspirations that parents have had for themselves. Because a large proportion of the respondents indicated that it was the prerogative of children to choose for themselves, those who had school-age children were asked whether their children had any job preferences. Responses (Table 40) were scattered in 1960 but the reported aspirations of Mexican-American children were lower than those of Anglos or Negroes. Although children's aspirations for themselves as reported by parents were not identical

TABLE 40. OCCUPATIONAL ASPIRATIONS OF CHILDREN, AS REPORTED BY PARENTS, 1960 AND 1961 (IN PERCENTAGES)

	1960			1961	
	Mexican-American	Negro	Anglo	Negro	Anglo
Doctor, nurse	6	19	9	18	12
Priest, minister, nun	2	1	5	1	5
Teacher	3	5	4	6	7
Engineer, scientist, electronics	2	0	4	3	4
Entertainment (sports, music, acting)	5	8	5	6	3
Other professional (managerial, proprietor, technician)	1	3	9	3	8
White collar, unspecified, clerical, any easy job	5	3	2	4	3
Skilled laborer, carpenter	6	8	5	7	3
Other job	5	3	8	2	4
Inapplicable, no school age children, no preference expressed	65	51	50	52	50
	100	101	101	102	99

to those selected for them by parents there were similarities; for example, the modal category for Negro parents and children was that of doctor or nurse. If responses considered as professional, technical, managerial, or proprietor are combined for 1960, Negroes and Anglos have the same total percentage at this level—36 percent, but Mexican-Americans have only 19 percent. There are differences within this large category but just how meaningful is difficult to say. Some over-all stability in response patterns is noticeable, particularly when Negro responses between 1960 and 1961 are compared.

Summary Educational and Occupational Scales

In 1960 a scale measuring educational and occupational aspirations was constructed and is shown in Table 41. Respondents with the highest levels of aspiration had the following characteristics: (1) desired children to have a college education, (2) would be satisfied only if their children had some college education or a college degree, (3) disagreed that financially it would be practically impossible to put their children through college, and (4) would like their children to go into a specific profession or into a profession in general. When the distribution for each group was dichotomized between scale scores 0 and 1-4, Anglos had significantly higher educational and

TABLE 41. EDUCATIONAL AND OCCUPATIONAL ASPIRATION SCALE FOR CHIL-
DREN, 1960 AND 1961 (IN PERCENTAGES)

Scale Type		Mexican-American	Negro	Anglo	Negro	Anglo
		1960			*1961*	
Low level of aspiration	0	67	35	21	16	5
	1	12	24	19	14	11
	2	8	18	15	19	13
	3	3	10	28	34	29
High level of aspiration	4	11	13	15	17	42
		101	100	98	100	100

1960: R = .8984; MR = .6675; Mexican-American/Anglo χ^2 = 105.58, 1 d.f., p < .001;
    Negro/Anglo χ^2 = 12.07, 1 d.f., p < .001; Mexican-American/Negro χ^2 = 49.66,
    1 d.f., p < .001.
1961: R = .8620; MR = .6425; Negro/Anglo χ^2 = 9.22, 1 d.f., p < .01.

aspiration scores than either Mexican-Americans or Negroes, and
Negroes had higher scores than Mexican-Americans.

By now the reader may be dismayed because we have spoken so
frequently of the relationship of socioeconomic status to level of
aspiration as well as other "class-related variables," but failed to
introduce analyses with appropriate controls. It should be repeated
that the interrelationship of numerous variables of this nature, with
pertinent controls, will be presented in a later chapter. For the
present, it should be said that with income controlled, the
relationship of race and ethnicity to level of aspiration for children
remained practically unchanged.

When level of aspiration for children and income were compared,
without controlling for race or ethnicity, there was a significant
difference between income categories and level of aspiration. When
race and ethnicity were controlled, the only significant difference
within the Mexican-Americans was between the high and low income
categories and this was barely significant with low income Mexican-
Americans having lower levels of aspiration than high income
Mexican-Americans. There were no significant differences within the
Negro sample. Within the Anglo sample, high and low income
persons had significantly different levels of aspiration for children
but then again the difference was relatively small.

When income was controlled, level of aspiration was significantly
different between Mexican-Americans and Anglos in every income
category. Mexican-American and Negro levels of aspiration for
children were significantly different within every income category.

Negro and Anglo levels of aspiration for children differed within income categories, Anglos being somewhat higher, but the differences were not statistically significant.

Thus, it would seem that race and ethnic differences in level of aspiration for children have some but not a significant degree of relationship to income. The extent to which occupational level is a determinant of level of aspiration and other variations in race and ethnic scale scores will be thoroughly considered in due time.

Although neither the 1960 nor 1961 scales had a coefficient of reproducibility that quite met minimum standards for consideration as a Guttman scale, their minimum coefficients of reproducibility were such that we have included both scales as summary measures.

Respondents with the highest levels of aspiration had the following characteristics in the 1961 scale: (1) desired to have their children go to college or beyond, (2) had a special job that they would like their children to go into, (3) would not be satisfied if their children got one level less education than he thinks they will get, and (4) thought that their children would go to college or beyond. Anglo aspirations for their children were significantly higher than those of the Negroes.

Conclusion

Probably one of the most surprising responses concerning the socialization process was that about 30 percent of the Anglos but less than 5 percent of the Mexican-Americans thought that religion should be taught to girls in the home. Although proper dating etiquette for girls was mentioned by almost 10 percent of the Mexican-Americans but only 2 percent of the Anglos, this response was inconsistent with the prevalent notion that socialization in the home places great stress on dating etiquette among Mexican-Americans. A variety of unexpected responses of Mexican-Americans revealed Anglos to be at least, if not more, oriented toward the home as a place of socialization than Mexican-Americans. We have suggested that some of the literature may be outdated or simply may represent what has been going on in the minds of Anglo social scientists rather than what is going on in the minds of Mexican-Americans. We must also remember that socioeconomic status or social class differences in child-rearing patterns and beliefs undoubtedly play a part in explaining what have appeared to be racial or ethnic differences. Mexican-Americans and Negroes viewed the school as a locus of socialization to a greater extent than did Anglos but at the same time Anglo educational or occupational aspirations for their children were higher than Negro, and Negro aspirations were higher than Mexican-Americans. Controlling for socioeconomic status reduces and even changes the rank-order of these findings (as reported in other studies and as will be shown in a later chapter in this volume), that is, they are partially explained by socioeconomic status differences.

To summarize, Anglos (without the introduction of controls for socioeconomic status) expected their children to go farther in school and in the world of work and would be less likely to be satisfied if goals set forth for their children were not realized.

NOTES

1. The best contemporary summary of the literature on socialization may be found in William H. Sewell, "Some Recent Developments in Socialization Theory and Research," The Annals of the American Academy of Political and Social Science, Philadelphia, Vol. 340, September, 1963, pp. 163-181. While very little of the literature has been concerned with socialization of the Negro child or adolescent, most of what has been done in the scientific study of socialization is cited in this article. Citations of the literature that are found in the remainder of this chapter will mostly deal with articles that are specific to socialization as it differs for race and ethnic groups and/or socioeconomic status categories.

2. The role of the home as a place of socialization, particularly the role of the father is described in detail in Arthur J. Rubel's Across the Tracks: Mexican-Americans in a Texas City, Austin: The University of Texas Press, 1966, pp. 60-67. Rubel places great emphasis on the home as a place in which the boy is taught the norms of Mexican-American society. Similarly, he describes the role of the home for the socialization of the female child but in spite of this one gains the impression that a more direct type of action is taken to insure proper socialization of the male than the female.

3. George M. Foster describes the traditional Mexican-American family and compadrazgo in his volume Tzintzuntzan: Mexican Peasants in a Changing World, Boston: Little, Brown & Co., 1967, pp. 55-85. He speaks of the problem that the ethnographer has when his informants believe themselves compelled to describe the ideal family and ideal compadrazgo behavior though they realize that there is a gap between the ideal and real behavior. While they describe the Mexican family with warmth and have an image of themselves as loving parents, they also point out the role of beatings as the only sure way in which to imprint moral lessons on young minds. Through beatings children are supposed to learn that they owe unquestioning obedience to their parents, especially to their

fathers. Rather than accept the traditional descriptions of the Mexican-American family and the possibility that it is changing as a consequence of the move to northern urban areas, it would be better to think of the family as already in the process of change and perhaps much closer to that of comparable socioeconomic status Anglos than might be expected. One Mexican-American father discussed the extent to which he beat his boys in order to teach them obedience, stating that he has "always beat them like animals but that they still stay out all night." For an additional description of the role of the father and the home in socialization of the son, and one that is particularly pertinent since the fieldwork was done in Hidalgo County in the valley, see William Madsen's *The Mexican-Americans of South Texas,* New York: Holt, Rinehart & Winston, 1964, Chapter 6, "The Family and Society," pp. 44-57. While Madsen emphasizes the role of the family, he, just as Rubel, places great emphasis on the *palomilla,* the loosely knit play group of Mexican-American youth.

4. Rubel devotes an entire chapter to the family, and makes the point that young Mexican-Americans find some qualities of Anglo courtship and marriage quite attractive. Since the family is obviously in a process of change before leaving Texas, one would hardly expect to find only traditional patterns of behavior in the northern industrial community. For an excellent discussion see Rubel, Chapter 3, "The Family," op. cit. Also see Norman D. Humphrey, "The Changing Structure of the Detroit Mexican family: Index for Acculturation," American Sociological Review, Vol. 9, No. 6, December, 1944, pp. 622-625.

5. It should be made clear at this point that when we speak of the traditional Mexican-American family we are not referring to the type of convoluted thinking about it criticized by Miguel Montiel in "The Social Science Myth of the Mexican-American family," El Grito, Vol. III, No. 4, Summer, 1970, pp. 56-63.

6. Perhaps the study most pertinent to our work in terms of differences in level of aspiration between Negroes and whites is that described by Aaron Antonovsky and Melvin J. Lerner, "Occupational Aspirations of Lower Class Negro and White Youth," Social Problems, Vol. 7, No. 2, Fall, 1959, pp. 132-138. This study in an upstate New York city found that Negroes with a comparably low socioeconomic background not only had as high a level of aspiration as did whites but even higher. The question was how did they come to have such a high level of aspiration. The authors concluded that Negro youths perceive education as a road to occupational achievement and that they chose the professions since it would be unrealistic to seek success in the skilled trades, small business outside the Negro community, or in corporate structures. The authors concluded that in spite of every kind of deprivation in their backgrounds they had an intense desire to succeed and to get as far removed from their backgrounds as possible.

7. S. M. Miller and Frank Riessman, "The Working Class Subculture: A New View," Social Problems, Vol. 19, No. 1, Summer, 1961, p. 90.

8. When Grebler, Moore and Guzman's respondents in San Antonio were asked, "In your opinion, what are the main things that children need to be taught in the schools today?", the most frequent response emphasized traditional roles such as discipline, obedience, respect, good manners, or religious training. This was the case for 36 percent of the medium income respondents in

San Antonio and 41 percent of the low income respondents. Another 44 percent of their respondents stated that basic school subject matters, skills, and technical skills related to jobs or occupations should be taught in the schools. Others emphasized social skills such as getting along with others, adapting to society, and so on. When asked, "Do you feel the same for both boys and girls?", there were no sex differences in educational goals. Grebler, Moore and Guzman, *The Mexican-American People,* New York: The Free Press, 1970, pp. 367-368.

9. Aside from G. Franklin Frazier's classic, *The Negro Family in the United States,* Chicago: University of Chicago Press, 1939, one should read St. Clair Drake and Horace R. Cayton's *Black Metropolis,* New York: Harper & Row, 1962, particularly Chapter 20, "Lower Class: Sex and Family," pp. 564-599. In a subsection of this chapter, Dependent Men and Forceful Women, Drake and Cayton point out that while the lower class Negro male has been stereotyped as "lazy," "shiftless," there is another conception of the Negro male as a "healthy buck who can stand the heat of the blast furnace." How the lower class Negro male's partial absorption into the economy has affected his family relationship is described for the Depression period but what Drake and Cayton have to say is just as applicable to a large portion of the Negro population today.

10. Although we are not particularly concerned by the fact that Mexican-American and Negro women are not exposed to the literature of child-rearing experts, the fact that they are not (and for the larger part regardless of educational level) suggests why some of our responses concerning the relationship of Mexican-American and Negro parents to their children differs from that of the Anglo relationship. While a portion of this difference and in some cases a large portion has been attributed to social class differences in the past, the organization of society and segregation of minority groups from the larger community may account for the difference as much, if not more than, so-called social class differences. See Zena Smith Blau, "Exposure to Child-Rearing Experts: A Structural Interpretation of Class-Color Differences," The American Journal of Sociology, Vol. LXIX, No. 6, pp. 596-608.

11. Alphonso Pinkney concludes that in spite of the inconsistent reports on differences between Negro and white families and social class in reference to child-rearing, that middle class black parents are more likely to be like middle class white parents in their child-rearing practices than like lower class black parents. See Alphonso Pinkney, *Black Americans,* Englewood Cliffs: Prentice-Hall, Inc., 1969, pp. 97-98. One is inclined to agree with him, however much the current literature does suggest social class changes in child-rearing practices, as does Martha Sturm White's "Social Class, Child Rearing Practices, and Child Behavior," American Sociological Review, Vol. 22, No. 6, December, 1957, pp. 704-712. Also see William S. Bennett, Jr., and Noel P. Gist, "Class and Family Influences on Student Aspirations," Social Forces, Vol. 43, No. 2, December, 1964, pp. 167-173 for an assessment of material influences relative to parental influence at lower class levels. For two earlier articles dealing with social class and child-rearing, see Allison Davis, "American Status Systems and the Socialization of the Child," American Sociological Review, Vol. 6, No. 3, June, 1941, pp. 245-354, and Allison Davis and Robert J. Havighurst, "Social Class and Color Differences in Child-Rearing," American Sociological Review, Vol. 11, No. 6, December, 1946, pp. 698-710.

12. We are mainly concerned about differences in level of aspiration for children in terms of education and occupation and how they may influence the behavior of the child in terms of his later absorption and integration into the larger society. Two chapters in David Gottlieb and Charles E. Ramsey, *Understanding Children of Poverty*, Chicago: Science Research Associates, 1967, Chapter 2, "The Family and Social Life of the Deprived Child," and Chapter 3, "School Performance and Deprivation," should be particularly helpful in terms of understanding the milieu in which the child acquires his educational and occupational levels of aspiration. Also see Ralph H. Turner, *The Social Context of Ambition*, San Francisco: Chandler Publishing Co., 1964, 269 pp.

13. Quite aside from any differences in level of aspiration that would be expected on a basis of the assumed subcultural differences among the migrants, differences in occupational aspirations would be expected simply on a basis of socioeconomic status of "social class differences in the groups." LaMar T. Empey, "Social Class and Occupational Aspiration: A Comparison of Absolute and Relative Measurement," American Sociological Review, Vol. 21, No. 6, December, 1956, pp. 703-709, found that the absolute occupational status aspiration among high school seniors from the middle and upper classes are significantly higher than those of seniors from the lower classes. He also found that lower class seniors preferred and anticipated having significantly higher occupational status than their fathers. Empey concluded that lower class youth do not limit their occupational aspirations to the class horizon in which they have been reared but that their aspirations are not as high as those from upper strata.

But differences in occupational choices may not be so simply interpreted as differences in level of aspiration. Negroes in Columbus, Ohio were asked to rate 65 jobs on the North-Hatt scale. Their ratings were not significantly different from those made by white persons in 1947. In fact, the similarity of the two sets of ratings was striking. But differences in occupational aspirations by Negroes may be based on differences in the level of aspiration or lack of differences in level of aspiration but different evaluations of the jobs. See Morgan C. Brown, "The Status of Jobs and Occupations as Evaluated by an Urban Negro Sample," American Sociological Review, Vol. 20, No. 5, October, 1955, pp. 561-566.

14. Only brief mention can be made here of the vast and increasing literature on level of aspiration for youth. More recently much attention has been focused on the nature of the school, the neighborhood, and peer groups which may influence the child during the process of socialization and the acquisition of educational and occupational aspirations. The case for parental influence has been sufficiently made. We would expect some relationship between parent's current aspirations and what they report their children's aspirations to be, and the actual aspirations of the children. A few examples of the literature follow: William H. Sewell, Archie O. Haller and Murray A. Straus, "Social Status and Educational and Occupational Aspiration," American Sociological Review, Vol. 22, No. 1, February, 1957, pp. 67-73; Alan B. Wilson, "Residential Segregation of Social Classes and Aspirations of High School Boys," American Sociological Review, Vol. 24, No. 6, December, 1959, pp. 836-845; William H. Sewell, "Community of Residence and College Plans," American Sociological Review, Vol. 29, No. 1, February, 1964, pp. 24-38; Harry J. Crockett, Jr., "Social Class,

Education, and Motive to Achieve in Differential Occupational Mobility," The Sociological Quarterly, Vol. 5, No. 3, Summer, 1964, pp. 231-242; Irving Krauss, "Sources of Educational Aspirations among Working-Class Youth," American Sociological Review, Vol. 29, No. 6, December, 1964, pp. 867-879; William H. Sewell and J. Michael Armer, "Neighborhood Context and College Plans," American Sociological Review, Vol. 31, No. 2, April, 1966, pp. 159-168; and William H. Sewell and Vimal P. Shah, "Social Class, Parental Encouragement, and Educational Aspirations," American Journal of Sociology, Vol. 73, No. 5, March, 1968, pp. 559-572.

15. William H. Sewell and Vimal P. Shah, "Socioeconomic Status, Intelligence, and the Attainment of Higher Education," Sociology of Education, Vol. 40, No. 1, Winter, 1967, pp. 1-2, and 22. For the influence of parents' education on children's aspirations, see William H. Sewell and Vimal P. Shah, "Parents' Education and Children's Educational Aspirations and Achievements," American Sociological Review, Vol. 33, No. 2, April 1968, pp. 191-209.

16. Although we have not questioned Mexican-American, Negro, or Anglo youth in respect to their educational aspirations, a variety of studies suggest that their aspirations probably do not differ from the aspirations of Anglos as much as do the aspirations of their parents for them differ from the aspirations of Anglo parents for their children of similar socioeconomic status. The Coleman Report and other studies are discussed at length as they apply to the Mexican-American case in Grebler, Moore and Guzman, op. cit., pp. 161-169.

17. Norval Glenn has surveyed Negro stratification studies published during the past 30 years and concluded that formal education has been the most important prestige criterion among Negroes, more important than among whites. He evaluates the propositions that Negroes may value education more than do whites because: (1) it is scarcer among Negroes, (2) it may be of greater utility to Negroes, and (3) Negroes may be more differentiated than whites in educational attainment. He discounts the importance of the scarcity argument since it is probably not much scarcer than income among Negroes. Of the three possible reasons for education having such prestige value among Negroes, Glenn concludes that degree of differentiation is probably the strongest reason for the higher prestige value of education among Negroes. Quite aside from his discussion of the reason for the prestige value of education, he shows that a total of 15 studies ranked it highest, followed by occupation and wealth or income. See Norval D. Glenn, "Negro Prestige Criteria: A Case Study in the Bases of Prestige," The American Journal of Sociology, Vol. LXVIII, No. 6, May, 1963, pp. 645-657.

18. Although the entire volume is pertinent to our discussion of the role of the family in socialization vs. the role of the school, particular reference should be made to Richard A. Cloward and James A. Jones's selection, "Social Class: Educational Attitudes and Participation," in Education in Depressed Areas, A. Harry Passow (ed.), New York: Teachers College Press, 1963, pp. 190-216. This paper is reporting on surveys conducted of attitudes of adult residents on the lower East Side of Manhattan by the New School of Social Work, Columbia, Univesity. Respondents were asked: "About how much schooling do you think most young men need these days to get along well in the world?" Respondents with children in school responded as follows: 81 percent of the middle class

stated that more than a high school education is required; 68 percent of the working class responded in the same fashion, as did 43 percent of the lower class. When respondents were separated on a basis of the occupational level that they would recommend young men aim for, there was a definite relationship between social class and the percentage saying that more than a high school education was required. For example, middle class persons suggesting that a youth aim for professional or semi-professional occupations stated that more than a high school education was required in 75 percent of the cases, while working class people did so in 62 percent of the cases, and lower class respondents in only 50 percent of the cases. This and a series of other questions indicated that knowledge about the requisite education for various occupational levels was unevenly distributed by social class. The authors concluded that lower class persons placed less emphasis on education for education's sake but tie it to level of occupational aspiration.

A disconcerting note consistent with what we have frequently stated may be found in Colin Greer, "Immigrants, Negroes and the Public Schools," The Urban Review, Vol. 3, No. 3, January, 1969, pp. 9-12. Greer states that public education has been a consequence of economic improvement but rarely the bootstrap to economic improvement. His historically oriented examination of the New York public school system concludes that too much faith has been placed upon education as a vehicle to facilitate absorption and integration of the urban Negro population.

19. Horacio Ulibarri has summarized his study of migrant workers in Arizona, Colorado, New Mexico, and Texas and their views of education as follows: "I want my children to get an education, so they won't have to work as hard as I have." The majority of the persons in his sample were doubtful that their children would finish high school. Furthermore, the children expressed the same attitude but neither parent nor child was concerned about the problem. See "Social and Attitudinal Characteristics of Spanish-Speaking Migrant and Ex-Migrant Workers in the Southwest," in Mexican-Americans in the United States, John H. Burma (ed.), Cambridge: Schenkman Publishing Company, 1970, p. 32.

20. When Wendling and Elliott asked "What are the most important reasons for completing high school?" and handed their respondents a card with six established categories, they found that 19 percent of the middle class selected "to get a good job" in contrast to 34 percent of the working class and 40 percent of the lower class. The percent who selected "to get into college" increased from 49 to 57 to 74 percent going from the lower class to the middle class. When broken down by race and ethnicity it was interesting to note that in every case, regardless of class, Negroes selected "to get into college" as a reason for going through high school with whites in second place and Mexicans least frequently giving this response. In essence, it seems that it makes little difference which study we pick; Negroes have high levels of educational and occupational aspirations when social class is controlled. Aubrey Wendling and Delbert S. Elliott, "Class and Race Differentials and Parental Aspirations and Expectations," Pacific Sociological Review, Vol. 11, No. 2, Fall, 1968, p. 131.

21. In reporting their findings, Wendling and Elliott, op. cit., pp. 123-152, first point out that enormous differences were found in response to the kinds of

questions put to mothers of the school children: (1) "How much further in school should——go before he or she stops and works full time?" (2) "Since things do not always work out the way we would like them to, how much school or education do you think——will actually get?" Seventy-four percent of the mothers in the middle class category compared to 48 percent in the working class and 34 percent in the lower class held college graduation aspirations for their children. But among the middle class, only 54 percent expected their children to graduate, 31 percent among the working class and 19 percent among the lower class respectively. Having reviewed the literature carefully, the authors were aware of the contradictory and somewhat confusing findings previously reported and acknowledged that social class differences in level of aspiration are probably confounded by race and ethnicity and other variables as well. When race and ethnicity alone were controlled, 52 percent of the whites, 44 percent of the Negroes, and 26 percent of the Mexican-Americans stated that they aspired to college graduation for their children. Expectations showed a similar trend with 34 percent of the whites, 29 percent of the Negroes, and 13 percent of the Mexican-Americans stating that they expected their children would complete college. While control of class resulted in a reversal of the findings, with Negroes placed ahead of the whites in both aspiration and expectations in the lower class, the discussion of the findings on this matter will be postponed until a later chapter where aspiration, world view, income, and so forth, are discussed in their interrelationships for the Racine data.

22. Wendling and Elliott, op. cit., pp. 128-129, also reported that even with social class controlled Negroes expressed higher occupational aspirations than did the Anglos or Mexican-Americans.

23. The manner in which middle class parents instill in their children a degree of self direction is discussed by Melvin L. Kohn, "Social Class and Parent-Child Relationships: An Interpretation," The American Journal of Sociology, Vol. LXVIII, No. 4, January, 1963, pp. 471-480.

24. Herbert H. Hyman, "The Value Systems of Different Classes: A Social Psychological Contribution to the Analysis of Stratification," in Class, Status, and Power, Reinhard Bendix and Seymour Martin Lipset (eds.), New York: The Free Press, pp. 426-442.

Chapter 9

WORLD VIEW

Introduction

The term *world view* has been used by sociologists and anthropologists to designate the central core of meaning in societies or in certain groups within any one society. This central core of meaning is derived from the life-ways which constitute the designs for living of a given community, tribe, region, ethnic group, religious group, or socioeconomic category, and from which most individuals within any of these social entities construct their definitions of situations or the perspective from which they view events in the world about them.[1]

Mexican-American Versus Anglo World View: 1959

Perception of the World

It was hypothesized that when inmigrants adopt the world view of the dominant culture it implies the internalization of a set of values which facilitate economic absorption.[2] Consequently, a series of questions and statements were posed to elicit direct responses from the respondents describing their perceptions of the world and of their success in coping with it. It was disappointing to find that the respondents could not readily understand all of the questions and propositions assumed to be the "heart" of the study. Despite these difficulties, the surprising result was the close relationship of world view to other variables.

The basic question in 1959 was: "Some people feel that their lives have worked out just about the way they wanted. Others feel that they've had bad breaks. How do you feel about the way your life is working out?" Eighty percent of the Mexican-Americans and the Anglos were entirely or predominantly satisfied and relatively few (17 percent) were entirely or predominantly dissatisfied. Dissatisfaction was attributed to bad luck, bad breaks, or fate by more of the Mexican-American males (11 percent) than by Anglo males (4 percent), residual evidence of a southwestern Mexican world view in which the vicissitudes of life are viewed as the result of impersonal mechanisms of fate.[3]

Life Goals

Respondents were asked: "What are the things that a man wants most out of life?" Both Mexican-American (48 percent) and Anglo (40 percent) males mentioned factors such as good pay, money, savings, and good jobs more often than the Mexican-American (10 percent) and Anglo (16 percent) females. The goals of 56 percent of the Mexican-American and 68 percent of the Anglo females were more family oriented—they desired a good spouse, a happy family, and many children. It is apparent that sex differences were significant but that ethnic differences were relatively unimportant. Since the male is somewhat more dominant in the decision-making process among Mexican-Americans than among Anglos, did male values influence family decisions in a different direction in the Mexican-American than in the Anglo family? In the decision-making process, did sex differences in world view translate into ethnic differences? The data showed that male-female differences within ethnic groups were statistically significant but some sex differences between ethnic groups were not.

When respondents were asked if "getting ahead" was one of the things a man wanted most out of life, Mexican-Americans (78 percent) and Anglos (70 percent) were in almost complete agreement. Responses to the question, "As you see it, what are the most important things you must do if you want to get ahead?" suggested a greater disparity on the basis of sex rather than ethnic differences. Males were more likely to be concerned with work, or work and saving, while females were interested in saving (no mention of work), education, and the development of good personal characteristics. Anglos and Mexican-Americans had one difference of opinion: more

Anglos (8 percent) than Mexican-Americans (one percent) mentioned the need to strive or to be ambitious in order to get ahead.

Time Orientation and Planning for the Future

Time orientation is an important, perhaps one of the most crucial, facets of world view. Traditionally, Mexican-Americans are believed to possess a strong orientation to the present rather than to the past or future. As a consequence, the Mexican-American neither plans for the future nor does he have much hope of manipulating his environment or of controlling his destiny beyond the level of day-to-day decision-making.[4] When Mexican-Americans and Anglos were asked if it were possible to plan ahead in life, 42 percent of the Mexican-Americans but only 27 percent of the Anglos responded in the most negative category. Ethnic differences were statistically significant when sex was held constant. Sex differences within ethnic groups were evident but not significant. Typical answers of a large proportion of the Mexican-American females, who stated that one cannot plan, were:

> ". . . I don't think you can plan, because you plan one thing and God has something else planned for you."

> "You can't plan, because we don't know if we'll all be alive tomorrow or what God has in store for us."

> "No one can plan, because we never can tell what lies ahead in the future. We do *think* about plans but sometimes they don't turn out. So it's just what lies ahead."

Developing the 1960 World View Scale

Three Components of the Scale

When it was decided to measure world view in 1960 with closed-ended questions we thought that the questions should be constructed so as to determine if the respondent had an individualistic, active view on the one hand or a group-oriented, passive view on the other. It became apparent in pretests that three facets of world view were involved—group values, temporal orientation, and manipulative power.[5] The definition of world view for this research thus became (1) a person's perception of his own manipulative power versus the organization of the society or some other more powerful determinant, (2) his time perspective as oriented toward the present versus the future, and (3) his hierarchy of values that places individual achievement against ties to the group. Eight world view statements were employed with the results shown in Table 42. A discussion of each precedes the Guttman scale developed from seven of the questions.[6]

The Questions and Responses

Mexican-Americans agreed with the statement, "The secret of happiness is not expecting too much and being content with what comes your way," to a significantly greater extent than either Anglos or Negroes. The Anglos expressed greater agreement with this attitude than the Negroes, an outcome inconsistent with the white stereotype of the Negro at or before 1960. Even as late as the 1960's, white incomprehension of Negroes' aspirations was marked—witness the inability of a large segment of the population to foresee the willingness of Negroes to resort to violent and nonviolent action to secure some recognition of their aspirations.

On the other hand, there was almost no difference between the samples in response to "When a man is born, the success he is going to have is not already in the cards; each makes his own fate." Eighty to 90 percent were in agreement that what happens is *not* "in the cards." The Mexican-American responses seemed inconsistent with expectations derived from the sociological and anthropological literature but the differences may be indicative of a degree of value assimilation or cultural integration.

Anglos disagreed to a significantly greater extent than Mexican-Americans or Negroes with statements about planning for the future: "Nowadays (1960) with world conditions the way they are, the wise person lives for today and lets tomorrow take care of itself"[7] and "Planning only makes a person unhappy since your plans hardly ever work out anyway." Anglos had a significantly more future-oriented perspective than did others, which may well be a consequence of their life expectations. They have found that their plans do materialize while Mexican-Americans and Negroes have discovered the opposite. Each subculture subscribes to a realistic word view based on past experience.[8]

Responses to "It's always a good idea to put away most of your money for a rainy day," markedly differentiated Anglos from Mexican-Americans and Negroes. The Anglos strongly disagreed that saving for the future was necessary. Perhaps they had learned that saving is not really as important in our society as it once was, or at least that saving is not crucial for those who have been integrated into the economy and are covered by various social insurance plans. By contrast, the Mexican-American migratory worker and the Negro farm laborer, both of whom are marginal to the economy, are not covered by social welfare plans, unemployment insurance, or workmen's compensation. Tragedy strikes them hard indeed and makes them aware not only of the need to save but more painfully of their inability to do so. The data showed Mexican-Americans in the greatest agreement with this statement, a reflection of their almost complete lack of employment security.

The Mexican-American culture, unlike the Anglo which on the whole is characterized by an individualistic orientation, is believed to place more stress on group orientation, that is, the will of God is accomplished by placing the family and friends above oneself.[9] Of course, the Mexican-American male may express an orientation toward the group but place himself first in certain real-life situations, just as the Anglo may profess a set of humanitarian beliefs but not apply them to day-to-day activities. There was considerable unanimity in every sample in response to "Not many things in life are worth the sacrifice of being away from your family," but there was less agreement when the reference was to one's friends. Mexican-Americans and Negroes valued their friends significantly more than Anglos but there was no significant difference between Mexican-Americans and Negroes.

The last statement, "The best job to have is one where you are part of a group all working together, even if you don't get much indivdual credit," did not reveal the usual difference between Mexican-American group orientation and Anglo individualism. The only significant difference appeared when more Negroes indicated a willingness to work with others than did Anglos.

Most of the reactions to each proposition were generally of the nature expected from each ethnic group though perhaps the differences were not as pronounced as anticipated from perusal of the literature on their value systems. We had hypothesized that the deep pervasive fatalism characteristic of traditional Old Mexico or of the American rural Southwest would systematically differentiate the Mexican-Americans from the Anglos or Negroes; this difference was simply not in evidence in our Mexican-American sample. Perhaps migration to an industrial community is in itself a self-selective

TABLE 42. WORLD VIEW OF RESPONDENTS, 1960 (IN PERCENTAGES)

1. "The secret of happiness is not expecting too much and being content with what comes your way."

	Strongly Agree	Agree	Pro-Con	Disagree	Strongly Disagree	Not Ascertained	Total
Mexican-American	2	87	3	5	1	1	99
Anglo	4	65	8	20	3	0	100
Negro	1	58	13	25	3	1	101

2. "When a man is born, the success he is going to have is not already in the cards; each makes his own fate."

	Strongly Agree	Agree	Pro-Con	Disagree	Strongly Disagree	Not Ascertained	Total
Mexican-American	4	75	4	10	0	6	99
Anglo	10	81	5	4	0	0	100
Negro	7	80	6	6	1	1	101

3. "Nowadays, with world conditions the way they are, the wise person lives for today and lets tomorrow take care of itself."

	Strongly Agree	Agree	Pro-Con	Disagree	Strongly Disagree	Not Ascertained	Total
Mexican-American	1	44	8	39	3	5	100
Anglo	2	12	4	74	8	0	100
Negro	3	42	4	45	5	1	100

4. "Planning only makes a person unhappy since your plans hardly ever work out anyway."

	Strongly Agree	Agree	Pro-Con	Disagree	Strongly Disagree	Not Ascertained	Total
Mexican-American	8	49	11	27	4	2	101
Anglo	4	14	6	65	12	0	101
Negro	4	47	4	41	3	1	100

5. "It's always a good idea to put away most of your money for a rainy day."

	Strongly Agree	Agree	Pro-Con	Disagree	Strongly Disagree	Not Ascertained	Total
Mexican-American	7	73	6	13	1	1	101
Anglo	2	24	11	57	6	0	100
Negro	3	60	11	24	1	0	99

TABLE 42 (Continued)

6. "Not many things in life are worth the sacrifice of being away from your family."

Mexican-American	14	69	5	12	0	0	100
Anglo	6	77	6	12	0	0	101
Negro	4	84	4	7	0	0	99

7. "Not many things in life are worth the sacrifice of being away from your friends."

Mexican-American	1	34	11	51	2	1	100
Anglo	1	17	7	69	6	0	100
Negro	1	29	9	60	1	1	101

8. "The best job to have is one where you are part of a group all working together, even if you don't get much individual credit."

Mexican-American	2	62	8	21	3	4	100
Anglo	2	57	8	30	3	0	100
Negro	4	66	8	20	2	1	101

Question 1: Mexican-American/Anglo χ^2 = 34.91, 1 d.f., p < .001; Negro/Anglo χ^2 = 6.30, 1 d.f., p < .02; Mexican-American/Negro χ^2 = 65.53, 1 d.f., p < .001.

Question 2: Mexican-American/Anglo χ^2 = 4.45, 1 d.f., p < .05; Negro/Anglo χ^2 = 1.09, 1 d.f., not significant; Mexican-American/Negro χ^2 = .98, 1 d.f., not significant.

Question 3: Mexican-American/Anglo χ^2 = 67.63, 1 d.f., p < .001; Negro/Anglo χ^2 = 66.09, 1 d.f., p < .001; Mexican-American/Negro χ^2 = .13, 1 d.f., not significant.

Question 4: Mexican-American/Anglo χ^2 = 90.59, 1 d.f., p < .001; Negro/Anglo χ^2 = 71.48, 1 d.f., p < .001; Mexican-American/Negro χ^2 = 1.94, 1 d.f., not significant.

Question 5: Mexican-American/Anglo χ^2 = 149.09, 1 d.f., p < .001; Negro/Anglo χ^2 = 78.11, 1 d.f., p < .001; Mexican-American/Negro χ^2 = 16.89, 1 d.f., p < .001.

Question 6: Mexican-American/Anglo χ^2 = .00, 1 d.f., not significant; Negro/Anglo χ^2 = 3.45, 1 d.f., not significant; Mexican-American/Negro χ^2 = 2.81, 1 d.f., not significant.

Question 7: Mexican-American/Anglo χ^2 = 18.99, 1 d.f., p < .001; Negro/Anglo χ^2 = 9.23, 1 d.f., p < .01; Mexican-American/Negro χ^2 = 1.83, 1 d.f., not significant.

Question 8: Mexican-American/Anglo χ^2 = 2.79, 1 d.f., not significant; Negro/Anglo χ^2 = 7.00, 1 d.f., p < .01; Mexican-American/Negro χ^2 = .54, 1 d.f., not significant.

process—those who migrate must not possess a fatalistic world view, as indicated by their positive attempt to improve their economic position.

The World View Scale

Seven of the above eight propositions were included in the World View Scale presented in Table 43. The statement about saving for a rainy day was eliminated because the pattern of its responses was so inconsistent with those of the other seven statements. The respondent who scored highest on the scale, that is, the person who was most passive and group oriented agreed that:

(1) Not many things in life are worth the sacrifice of being away from your family.
(2) The secret of happiness is not expecting too much and being content with what comes your way.
(3) The best job to have is one where you are part of a group all working together, even if you don't get much individual credit.
(4) Planning only makes a person unhappy, since your plans hardly ever work out anyway.
(5) Nowadays with world conditions the way they are, the wise person lives for today and lets tomorrow take care of itself.
(6) Not many things in life are worth the sacrifice of moving away from your friends.

and disagreed with:

(7) When a man is born the success he is going to have is not already in the cards—each makes his own fate.

TABLE 43. WORLD VIEW SCALE, 1960 (IN PERCENTAGES)

	Scale Type	Mexican-American	Anglo	Negro
Individualistic and Active	0	2	6	3
	1	3	12	11
	2	13	22	8
	3	19	37	22
	4	13	7	8
	5	19	11	25
	6	27	6	21
Group Oriented and Passive	7	4	0	2
		100	101	100

R = .9001; MR = .7125; Mexican-American χ^2 = 91.74, 1 d.f., p < .001; Negro/Anglo χ^2 = 65.14, 1 d.f., p < .001; Mexican-American/Negro χ^2 = 2.36, 1 d.f., not significant.

When scale scores were dichotomized (using scores 0-3 and 4-7 as the two halves of the dichotomy), Anglos were significantly more individualistic and less passive than Mexican-Americans or Negroes, but the world view scores of Mexican-Americans were not significantly different from those of the Negroes.

A Methodological Concern

The Consequence of Meaningless Questions

At this juncture we are concerned not with whether the world view statements were the most representative of the universe of questions that could have been asked or whether they were the "best" questions to ask at this point in the life cycle of respondents, but with whether the questions were as meaningful to the respondents in the Racine samples as they were to persons in the initial pretests in another community of more acculturated Mexican-Americans. If the world view propositions were not significant to a portion of the respondents, should that portion have been removed from the sample prior to an analysis of the world view responses? If the views of a substantial portion of the Negro and Mexican-American respondents were not properly represented by their replies to the world view questions, might not elimination of respondents decrease or perhaps eliminate differences between Mexican-Americans, Negroes, and Anglos? If only those who found the questions meaningful were included in the analysis, would the relationship between world view and other variables be changed?

To include in the analysis only those who appeared to find the questions meaningful might present a biased set of findings. Those permitted to remain in the sample would be only those who had been integrated into the larger society to the extent that they responded to the world view questions in essentially the fashion expected by their integrated counterparts (Mexican-American and Negro interviewers who were second and third generation residents in an urban, industrial community in northern Illinois). With this technique we would be removing those who were likely to have been absorbed into the economic system at a lower level and less integrated into the larger society. Theoretically, the reduced samples would be more homogeneous than the original; they would preclude the possibility of describing the relationship of antecedent and intervening explanatory variables to the full range of absorption and integration. On the other hand, if difficulty with the world view questions was experienced on a random basis among the respondents, then the concerns that have just been expressed would not be operative; essentially the same world views and relationships would be found in the smaller sample as were found in the larger.

With the foregoing discussion in mind, 265 respondents who did not appear to comprehend the world view questions, either through confusion or failure to answer, were eliminated from the larger sample.[10] The criteria for omission was an evaluation by the interviewers and not by those who were conducting the analysis.[11] Mexican-Americans and Negroes were eliminated disproportionately to Anglos but their selection was based on how Mexican-American or Negro interviewers (not Anglo) reacted to their responses.

Before proceeding we must decide whether the elimination of 265 respondents changed the original differences between groups. Were those who were eliminated randomly distributed or were they the least absorbed and least integrated of each group?

A Comparison of the Original Sample and the World View Sample

The original 1960 sample consisting of 236 Mexican-Americans, 284 Anglos, and 280 Negroes is designated in Table 44 as S1. The subsample includes 164 Mexican-Americans, 163 Negroes, and 218 Anglos. It will be referred to as S2.

In the sample of 800 and in the reduced sample of 545, Anglos had significantly more active world view scores than Negroes; the latter, although more active in world view than Mexican-Americans,

TABLE 44. TOTAL SAMPLE VERSUS WORLD VIEW SAMPLE, 1960 (IN PERCENTAGES)

Scale Type		Mexican-American		Negro		Anglo	
		S1	S2	S1	S2	S1	S2
Individualistic and Active	0 – 3	37	39	44	48	77	81
Group-Oriented and Passive	4 – 7	63	61	56	52	23	19
		100	100	100	100	100	100

Total Sample (S1): N = 800; Mexican-American = 236, Negro = 280, Anglo = 284.
World View Sample (S2): N = 545; Mexican-American = 164, Negro = 163, Anglo = 218.

were not significantly so.[12] Removing confused respondents and those who failed to answer the questions did not result in significant change in the pattern of world view scores by race or ethnicity. The differences between the samples which did occur were in the direction expected. While group-oriented, or passive respondents were disproportionately eliminated from the original race and ethnic samples, the correspondingly reduced samples were only 3 or 4 percent more individualistic or active than the original samples.

Further tests were undertaken to increase our assurance that the composition of S1 was not significantly different from that of S2. There was no significant difference in the level of aspiration for children scale scores between the original sample of 800 and the reduced world view sample of 545 taken either in total or by individual racial or ethnic subgroups.[13] Nor did vocational aspirations for children differ significantly between S1 and S2.[14] Education was dichotomized at a meaningful cutting point for urban, industrial persons with backgrounds similar to those of our respondents—better educated respondents were defined as those with nine or more years of education. Although the two Mexican-American samples (original and reduced) were almost identical in all respects, there were fewer less educated Negroes and Anglos in the sample of 545 than in the original. The difference was statistically significant for the Anglo sample.[15] While it might be surprising that less educated Mexican-Americans were not eliminated from the sample of 545, this fact plus illuminating incidents in the interviews made it apparent that years of formal education had little influence in determining whether or not a question was meaningful to a Mexican-American respondent. Educated or not, some Mexican-Americans laughed at the questions and stated that the world view questions were "silly." By contrast, those who were eliminated from the Negro and Anglo samples were persons with less education, creating at least the presumption that the questions were more meaningful to, or at least more readily answered by, better educated Negroes and Anglos. When S1 and S2 were compared in terms of their income distribution, no significant differences were found between the combined samples or among any of the three racial and ethnic groups.[16] And contrary to expectations, there were no significant differences between S1 and S2 on a basis of time in the community.[17]

It can be concluded that those who were eliminated from the larger sample on a basis of their confusion or inability to respond to the world view questions were eliminated on a more or less random basis. Less absorbed and less integrated Mexican-Americans and Negroes were not systematically excluded by the process.[18]

The Relationship of World View to Race or Ethnicity, Religion, and Sex

It was initially hypothesized that world view of respondents would vary individually with race or ethnicity, religious affiliation or identification, and sex. As a consequence of the combined effect of sex, religion, and race or ethnicity, Catholic Mexican-American females and Protestant Anglo males were expected to be at opposite ends of the world view scale.[19] Instead, the data showed, as controls were introduced, that significant differences were related to race or ethnicity and to sex, but that there was relatively little correlation between religious affiliation and world view. The significant differences that appeared between the Mexican-American sample, which was predominantly Catholic, the Negro sample, which was predominantly Protestant, and the Anglo sample, religiously heterogeneous, did not support the hypothesis of a relationship between religious affiliation and world view, but showed merely that variations between the Mexican-American, Negro, and Anglo samples were slightly increased by religious composition.[20] It was, however, when race or ethnicity was combined with religion and sex that the differences in world view were greatest and most significant.

Perhaps the best way to summarize the relationship of world view to race or ethnicity, religion, and sex is to create a table in which each group is dichotomized at the same point on the world view scale as either independent and active or group-oriented and passive. This makes it possible to compare each race or ethnic group, and the males and females within each ethnic group. Table 45 simply shows

TABLE 45. PERCENT PASSIVE AND GROUP-ORIENTED ON WORLD VIEW SCALE, 1960 (IN PERCENTAGES)

Mexican-American 63				Negro 56		Anglo 23			
Catholic 63		Protestant 63		Protestant 56		Catholic 27		Protestant 21	
Female	Male	Female	Male	Female	Male	Female	Male	Female	Male
70	55	71	55	60	52	36	18	26	16

the percentage of each group that may be characterized as passive and group-oriented on a basis of their responses to the questions on the world view scale.

Mexican-Americans were significantly more passive in their world views than Anglos but not significantly more so than Negroes, although Negroes were significantly more passive than Anglos. There was relatively little difference in world view between Catholic and Protestant Mexican-Americans, but we must add that there were relatively few Protestants. Catholic Anglos were more passive than Protestant Anglos but the difference was not significant. Mexican-American females, both Catholic and Protestant, were significantly more passive than were the males. Negro females were more passive than males, but not significantly so. Females were significantly more passive than males among the Catholic Anglos but Anglo Protestant females, although more passive than males, were not significantly so. The greatest significant difference among ethnic/religious groups was between the Mexican-American Catholics and Anglo Protestants. Within Protestant groups the greatest variation was between Anglo males and Negro females. When sex was controlled, Mexican-American Catholic females differed most from Anglo Protestant females. When cross ethnic/religious/sex comparisons were made, the greatest difference was between Mexican-American Catholic females and Anglo Protestant males, exactly as hypothesized. Examination of every possible combination of race or ethnicity, religion, and sex led to the conclusion that race or ethnicity had the greatest effect on world view, with sex next, and religion the least. When the effects of the three variables were combined, then differences between groups were maximized, as shown by the data in Table 45.[21]

The Relationship of Selected Variables by Race or Ethnicity, Religion, and Sex to World View

A Recapitulation of the Variables

Before final presentation of the interrelationship of variables with world view (Table 46), a recapitulation of the relationship of each of these variables to race or ethnicity, religion, and sex (Table 46) might be helpful, albeit a bit repetitious. Some of the greatest disparities to be found between groups were those based on education. Anglos were by far the best educated, with the Negroes second, and the

TABLE 46. RELATIONSHIP OF SELECTED VARIABLES TO RACE OR ETHNICITY, RELIGION AND SEX: EACH VARIABLE DICHOTOMIZED (DISTRIBUTION IN PERCENTAGES)

	WORLD VIEW		EDUCATIONAL LEVEL		LEVEL OF ASPIRATION FOR CHILDREN		LEVEL OF LIVING		OCCUPATIONAL LEVEL	
	Active	Passive	8 Yrs. or Less	9 Yrs. or More	Low	High	Low	High	Low	High
MEXICAN-AMERICAN	37	63	83	17	79	21	68	32	84	16
CATHOLIC	37	63	83	17	80	20	67	33	83	17
Cath. Male	45	55	83	17	70	30	73	27	82	18
Cath. Female	30	70	82	18	88	12	60	40	85	15
PROTESTANT	37	63	88	12	71	29	82	18	88	12
Prot. Male	45	55	89	11	78	22	78	22	78	22
Prot. Female	29	71	87	13	63	37	88	12	100	0
NEGRO	44	56	43	57	59	41	73	27	73	27
PROTESTANT	44	56	43	57	59	41	73	27	73	27
Prot. Male	48	52	50	50	58	42	72	28	72	28
Prot. Female	40	60	37	63	60	40	73	27	75	25
ANGLO	77	23	17	83	42	58	11	89	30	70
CATHOLIC	73	27	15	85	40	60	13	87	30	70
Cath. Male	82	18	20	80	46	54	15	85	25	75
Cath. Female	64	36	11	89	34	66	11	89	34	66
PROTESTANT	79	21	18	82	44	56	9	91	30	70
Prot. Male	84	16	24	76	42	58	11	89	32	68
Prot. Female	74	26	12	88	47	53	6	94	29	71

Mexican-Americans last, with less difference between Anglos and Negroes than between Negroes and Mexican-Americans. Eighty-three percent of the Anglos had nine or *more* years of education while 83 percent of the Mexican-Americans had eight years of education or *less.* The difference was statistically significant at the .001 level between each of these groups.

There was no significant difference in educational level of Catholics or Protestants within either the Mexican-American or the Anglo samples. Anglo males, both Catholic and Protestant, had less education than the females but not significantly so. Among the Negroes, males had significantly less education than females.

The largest contrast in education between Mexican-Americans and Anglos was between females, Catholic or Protestant, within religious groups, or across religious groups. Almost as great was the difference between Mexican-American Catholic males and Anglo Protestant females, and between Mexican-American Catholic females and Anglo Catholic females.[22]

It will be recalled that Anglos had the highest level of aspiration for their children, followed by the Negroes, with the Mexican-Americans at the lowest point; all differences were statistically significant, but did not compare in size with differences based on education. The Mexican-American Catholics had somewhat lower levels of aspiration for their children than did the Mexican-American Protestants; in the Anglo sample the Protestants had lower levels of aspiration for their children than the Catholics, but not significantly so. The difference between Mexican-American and Anglo levels of aspiration appeared greatest among Catholics, even greater than that between Mexican-American Catholics and Anglo Protestants. The Mexican-American Catholic females had significantly lower levels of aspiration for their children than did the males; Mexican-American Protestant males had lower levels of aspiration than did the females. Within the Anglo Protestant and Catholic groups the difference on a basis of sex was the opposite of that among the Mexican-Americans, but not statistically significant—Anglo Catholic males and Protestant females had the lowest levels of aspiration for their children. Among the Negroes, females had only slightly lower levels of aspiration for their children than did males.

The greatest difference between groups on level of aspiration for children was between the high-aspiring Anglo Catholic females and the low-aspiring Mexican-American Catholic females. It is very clear that this startling contrast influenced every Mexican-American and Anglo comparison, including that of Catholics and females.

The level of living (possessions) scale was the only scale on which Mexican-Americans and Anglos were not at opposite extremes. Anglos had the highest level of living, followed by the Mexican-Americans, and then Negroes, with little difference between the Mexican-Americans and Negroes. Among the Mexican-Americans, Catholics had a higher level of living than Protestants, whereas among the Anglos, Protestants were higher than Catholics, but neither difference was significant. The highest level of living scores were obtained in homes where Anglo Protestant females were interviewed and the lowest level of living scores were obtained in Mexican-American Protestant female interviews.

Anglos were at the highest occupational levels, with Negroes next, and Mexican-Americans at the lowest level. Each group was significantly higher than the other but with greater differences between Anglos and Negroes than between Negroes and Mexican-Americans. Within the Mexican-American group, Catholics and Protestants had similar levels of living but the Catholics were somewhat higher. Catholics and Protestants had almost identical occupational levels among the Anglos.

Complexity as a Consequence of Sophistication

There are basically two alternative approaches to explaining the relation of world view to other variables. World view could be regarded as the determinant of how much education a person will manage to obtain, the level of aspiration that he will have for his children, the occupational level at which he will be employed, and the level at which he will be living. Equally plausible is that years of education, occupational level, level of living, and the level of aspiration for children (all in interaction within a social context) will shape a person's total view of the world.[23] In either case, education is closely related to world view.

We hypothesized that Catholic influence on world view, as it has been traditionally perceived and especially in reference to the world view of Mexican-Americans, would be more likely than Protestant influence to overshadow the effect of education. More succinctly, traditional Catholic influence would take a different direction than education while Protestant influence would be in the same direction as education.[24] But, if it is in fact a combination of Catholicism and Mexican traditions rather than one or the other that is the determinant of the world view of the Mexican-American, then Anglo Catholics and Protestants with similar levels of education should have world views that differ less from each other than from the world views of Catholic Mexican-Americans.

If the Protestant Negro subculture emphasizes rewards at some future date rather than at the present, just as the traditional Mexican-American Catholic subculture, then the correlation between education and world view among Protestant Negroes would be more similar to that of Catholic Mexican-Americans than to either Protestant or Catholic Anglos.

Within this framework we could expect that some who have been well educated will in the process have acquired a world view that serves as a rationalization for their success in life—a world view which emphasizes personal effort as the key to success. Theirs would be an active, individualistic world view and one which would also include the notion that others who make the effort can be equally successful in manipulating their environment.[25] On the other hand, a highly sophisticated Anglo might contend that his position at the upper levels of society could be explained by the way that the system works—that it is more or less initial advantage and/or lucky breaks that have enabled him to move upward in the system.

We had hypothesized that the more educated persons within each religious group would have an active, independent world view and the less educated a more passive view. Parenthetically, we must insert that the world view of the educated probably does not "square" in all respects with how things really work in all segments of society. While a middle or upper socioeconomic status Anglo may well be able to manipulate his social environment to his advantage, not every educated person who has acquired an active world view is in a position to do so, particularly if he belongs to a minority racial or ethnic group.[26]

Those who have high aspirations but relatively little in the way of material goods, status, and education may rationalize their general lack by blaming it on how the system works or on external forces which determine what one will have in life. The less educated Mexican-American, for example, may wish much for his children but be willing to settle for less because the forces that determine how much education a child will receive and how far a child will move upward occupationally are defined as insurmountable or almost entirely beyond his control. For some, these forces are believed to be supernatural while others regard them as simply fortuitous or chance. The less educated Negro may share the same more or less fatalistic view but his explanation is based on the manner in which the white power structure prevents upward movement for some but facilitates upward movement for others, principally Anglos.[27]

Thus, it is possible for Anglos, Mexican-Americans, and Negroes with somewhat different world views to have the same score on the world view scale. Some believe that society is organized in such a way as to make it possible for certain categories of persons to achieve more readily than others; others believe that chance or powerful external forces of a mystical nature may dictate their lives. Both tend to be passive and fatalistic as a consequence.

Since Protestant Anglos and Catholic Mexican-Americans have tended to be at opposite extremes in most measurements of the variables and since there are few "sophisticated" Anglos in the sample, we would expect relatively high correlations between world view and other variables, particularly education and occupational level, for Protestant Anglos. The pattern of correlations for unsophisticated Catholic Anglos should be similar to that for Protestant Anglos. Protestant Negroes, on the assumption that some have a passive world view related to traditional patterns of socialization and education while others have a passive world view for essentially the same reason as the sophisticated Anglos, should have the next lowest set of correlations. Mexican-American correlations should be lowest of any ethnic group if we accept the Catholic Mexican-American background as the determinant of their world view rather than education, occupational level, or other variables.

There is also the possibility that none of the Anglo correlations will be very high if one kind of Anglo cancels out another or if world view is relatively independent of the variables with which we have compared it. Let us turn to the data presented in Table 47.

World View and Education

When the samples of Mexican-Americans, Anglos, and Negroes were combined there was a statistically significant correlation of .3531 between world view and education. In a 2 x 2 table, 57 percent of the Mexican-Americans were in the low education and passive world view cell, and 69 percent of the Anglos were in the opposite high education and active world view cell. The correlations

TABLE 47. RELATIONSHIP OF WORLD VIEW TO EDUCATIONAL LEVEL, LEVEL OF ASPIRATION FOR CHILDREN, OCCUPATIONAL LEVEL AND LEVEL OF LIVING, CONTROLLING FOR RACE OR ETHNICITY, SEX AND RELIGION

r_4 Coefficients of Correlation of World View and Selected Variables

	Educational Level	Level of Aspiration for Children	Occupational Level	Level of Living
Mexican-American				
Catholic M	.2324*	.1802†	.0610	.2921**
Catholic F	.2320††	.1959†	.0788	.1043
Negro				
Protestant M	.1693†	.1395	.0977	.2520**
Protestant F	.1374	.1904††	.1517	.1048
Anglo				
Catholic M	.4187*	.1402	.3780*	.1713
Catholic F	.1950	.0355	.0424	.2337
Protestant M	.3892**	.1967	.2478†	.1006
Protestant F	.2473†	.1828	.2126	.1457

r_4 coefficients of correlation significant at levels as indicated below:

† = χ^2 significant at .10 level
†† = χ^2 significant at .05 level
* = χ^2 significant at .02 level
** = χ^2 significant at .01 level

between world view and education were .3040 for Anglos, .1875 for Negroes, and .2423 for Mexican-Americans.

After introducing varying controls for race and ethnicity, religion, and sex, the Anglo Protestant males had a correlation of .3892 between world view and education, Anglo Catholic males .4187, Anglo Protestant females .2473, and Anglo Catholic females .1950, all but the last being statistically significant. Mexican-American Catholic females and males had essentially the same correlations between education and world view as did Anglo Protestant females. But none of the correlations which were statistically significant were sufficiently high to account for much of the variation in world view. We must conclude that despite differences between Anglos, Negroes, and Mexican-Americans in education and world views, the interrelationship of these variables within each ethnic group, religious group, and sex category gives little support to the more detailed subcultural hypothesis presented in pages 217 to 219.

World View and Level of Aspiration for Children

Level of aspiration for children was correlated with world view in the total sample in the same manner as education. The correlation was .2240, for Anglos .1287, Negroes .1657, and Mexican-Americans .2249; all were statistically significant. About half the Mexican-Americans had passive world views and low levels of aspiration for their children, and about half the Anglos had active world views and high levels of aspiration for their children. More of the Negroes were in the passive world view and high aspirations for children cell in proportion to Anglos or Mexican-Americans, suggesting that Negroes may have high aspirations for their children even though their view of the world is considerably different from that which is held by Anglos with high aspirations for their children.

When the Anglos were partialled by religion and sex, Protestant correlations were higher than the Catholic, but in no case was level of aspiration for children significantly related to world view. The best explanation for these low correlations is that there were a number of Anglos in the active world view and low level of aspiration for children cell who did not have specific occupational levels for their children but simply stated that they would "leave it up to the children." Because they did not express their high aspirations for children in precisely the same manner as did the Mexican-Americans and Negroes (though they may have meant that they would leave it up to their children if they selected an *appropriate* occupational level), the number of Anglos who potentially would be in the active world view and high aspiration for children cells was reduced and the number in the active world view and low aspirations for children was increased.

Although the relationships between world view and level of aspiration for children among the Mexican-Americans were statistically significant, the magnitude of these correlations was essentially the same as for Anglo Protestants. The Negro correlations were similar to the other groups except that the female correlation was higher than the male and was statistically significant.

World View and Occupational Level

Turning to the relationship of occupational level to world view, the total combined sample had a statistically significant correlation of .2207, partly because 70 percent of the Anglos were in the active world view and high occupational level cell with a correlation of .1395 and about 40 percent of the Mexican-Americans were in the opposing passive world view and low occupational level cell with a correlation of .1362. There was practically no relationship between world view and occupational level for the Negroes.

Both Catholic and Protestant Anglo males had world views that were significantly related to occupational level and in the direction hypothesized, i.e., that an active world view was associated with high occupational level. Other groups showed either less or in some cases practically no relationship between world view and occupational level.

The data suggest that the Anglo who has been absorbed into the economy at a relatively high level not only has a different world view than does the Negro and Mexican-American but also that world view among Anglo males varies directly with the degree of success in the world of work.

World View and Level of Living

The correlation between level of living and world view for the total combined sample was .2972. Seventy percent of the Anglos fell in the high level of living and active world view cell while 46 percent of the Mexican-Americans and 43 percent of the Negroes were in the low level of living and passive world view cell. The Anglo correlation was .1586, the Negro was .1572, and the Mexican-American .0988, the latter not statistically significant.

When all controls were introduced, the only significant correlations found between level of living and world view were for Negro Protestant males and Mexican-American Catholic males, although all others were at least .1000 and ranged upward to .2337.

Summary

Previous analysis of the data had already shown that race or ethnicity was an important determinant of the social space that a migrant occupied in his community of origin and in the host community. While the existence of more or less distinctive subcultures for Mexican-Americans and Negroes was demonstrated by different response patterns for samples of each of these groups as compared to Anglos and by differences in patterns of the interrelationship of the variables in a correlation matrix for each of the groups, race and ethnicity did not account for all variation in responses to the question. Although income and other characteristics of the respondents were related to world view, more variation within the total sample was explained by race or ethnicity than by any other variable.

Continued investigation raised the question of whether or not the addition of religion and sex as variables for control might not result in separation of the larger sample into yet more diverse subcultural groups with characteristics sufficiently distinctive to affect absorption and integration. Might not the combination of being a Catholic, being socialized in the Mexican-American tradition, and being female generate maximum differences in world view from that of the Anglo Protestant male? Would the intercorrelations of world view with the other variables differ within samples on any basis that could be predicted from existent theory about the interrelationship of world view, education, and measures of position within the larger society and its subgroups?

To answer the first question, Anglo Protestant males were at the extreme active end of the world view scale and Mexican-American Catholic females were at the passive end of the scale, accompanied by Mexican-American Protestant females. Although Anglo Catholics and Protestants were at the most favorable end of the scale and Mexican-American Catholics and Protestants were at the lowest end of the world view scale, neither religious nor sex differences presented such a systematic pattern for the other variables.

As for the second question, world view was significantly correlated (but none of the correlations was very high) with three of the four variables for one group only—the Mexican-American Catholic males. Over-all, the correlations of the greatest magnitude were for Anglo Catholic males; the lowest correlations were for Anglo Catholic and Negro Protestant females. Therefore, the hypothesis that the highest correlations between world view and the other variables, with controls for race and ethnicity, would be in the Anglo group (Protestant males) and the lowest correlations in the Mexican-American group (Catholic females), with other groups falling between the extremes in a systematic pattern, must be rejected. (It could be noted that Anglo Catholic and Protestant males had the highest correlations between world view and education or occupation but if we were to continue this "male" pattern, Mexican-Americans and Negroes would fall in an order which is also incompatible with the hypothesis.)

The general theory of subcultures differentiated by the interrelationships of variables is not rejected but the specific pattern of difference hypothesized was not present in the samples observed. If each of the variables is related to race or ethnicity, but does not vary with religion or sex in the systematic fashion suggested, or in any other readily discernible pattern, is it possible that these conjoint relationships do not really increase the visibility of various inmigrant groups?

NOTES

1. Florence R. Kluckhohn and Fred L. Strodtbeck, *Variations in Value Orientation*, New York: Row, Peterson & Co., 1961, p. 2.

2. Lest the reader confuse our concept of world view with a measure of what has become popularly known as the culture of poverty, a word should be said about the latter. By now, Michael Harrington's *The Other America* and Oscar Lewis's *Five Families, The Children of Sanchez*, and *La Vida* are as well known to laymen as to sociologists and anthropologists. Lewis and others treat the culture of poverty as a cross-cultural phenomena to be found in all ghettos. They describe family structure, masculine complexes, relationships with neighbors, and a view of the world, all of which supposedly differentiate persons in the culture of poverty from the remainder of the society. The "culture of poverty" is a design for living passed down from generation to generation through what might be called the mechanism of social inheritance. The danger here is that the concept may then be turned around to provide a rationale for the belief that the poor have only themselves to blame for their condition. The culture of poverty is erroneously held to be an explanation for the conditions under which the poor live but, as we see it, the living conditions of the poor should be considered an explanation of how a set of behavior patterns and a certain outlook on life are generated. Charles Valentine has been particularly critical of this view in his volume, *Culture of Poverty*. Perhaps the best discussion of the entire controversy may be found in Ulf Hannerz's *Soulside*, New York: Columbia University Press, 1969, in Chapter 9, "Mainstream in Ghetto Culture," pp. 177-192. Also see Section 4, "The Culture of Poverty," in *The Affluent Society*, Hanna H. Meissner (ed.), New York: Harper & Row, 1966, pp. 91-125.

3. As William Madsen has stated in *The Mexican-Americans of South Texas*. New York: Holt, Rinehart & Winston, 1964, Chapter 3, "La Raza," p. 16, "The Latin world view contains unpondered conflicts in the concepts of Divine will, individual will, and fatalism. Sometimes, events are explained as the result of the impersonal mechanism of fate, 'What will be, will be.' Certain individuals maintain that human fate is correlated in a mechanical way with the position of the heavenly bodies at the moment of birth. Others who accept the possibility of astrological divination point out that the course of the stars and planets is controlled by God. Most Latins believe that fate is a mechanism of God's will. Although the fate of the individual is decided before birth, God has the power to

alter it. Through prayer, sacrifice, and even bartering, one can induce God to modify one's fate. The paramount nature of Divine will is reflected in the saying *Haga uno lo que haga, todo es lo que Dios quiere* [Do what one will, everything is as God wishes]." And as Madsen continues, "Unlike the Anglo world view where man emerges as the dominant force except on Sunday, the Latin view conceives of God as all-powerful and man as but a part of nature that is subject to His will."

4. Madsen, op. cit., p. 17. In reference to planning for the future, Madsen states, "Acceptance and appreciation of things as they are constitute primary values of *La Raza*. Because God, rather than man, is viewed as controlling events, the Latin lacks the future orientation of the Anglo and his passion for planning ahead. Many Mexican-Americans would consider it presumptive to try to plan for tomorrow because human beings are merely servants of God and it is He who plans the future. The Latin lives for today instead of creating a blueprint for the future. He is dedicated to living the moment to its fullest in the roles assigned to him by God."

5. Although Florence Kluckhohn's work on value orientation has been helpful, we have been particularly interested in three of the orientations taken from Kluckhohn by Bernard C. Rosen in "The Achievement Syndrome: A Psychocultural Dimension of Social Stratification," American Sociological Review, Vol. 21, No. 2, April, 1956, pp. 203-211. Rosen mentions the following three dimensions: (1) what is the accepted approach to the problem of mastering one's physical and social environment, (2) what is the significant time dimension, and (3) what is the dominant modality of the relationship of the individual to his kin. Rosen's definition of the activist versus pacifist orientation is quite suitable for our use. "Activist-pacifist orientation concerns the extent to which a society or sub-group encourages the individual to believe in the possibility of manipulating the physical and social environment to his advantage. In an activist society the individual is encouraged to believe that it is both possible and necessary for him to improve his status, whereas a pacifistic orientation promotes the acceptance of the notion that individual efforts to achieve mobility are relatively futile." He describes present-future orientation: "Present-future orientation concerns a society's attitude toward time and its impact upon behavior. A present-oriented society stresses the merit of living in the present with an emphasis upon immediate gratifications; a future-oriented society urges the individual to believe that planning and present sacrifices are worthwhile, or morally obligatory, in order to insure future gains."

His familistic-individualistic orientation is also well stated: "Familistic-individualistic orientation concerns the relationship of the individual to his kin. One aspect of this orientation is the importance given to the maintenance of physical proximity to the family of orientation. In a familistically-oriented society the individual is not urged, or perhaps not permitted, to acquire independence of family ties. An individualistically-oriented society does not expect the individual to maintain the kinds of affective ties which will impede his mobility."

6. Six of our eight world view questions were exactly as they appeared in Bernard C. Rosen's "Race, Ethnicity, and the Achievement Syndrome," American Sociological Review, Vol. 24, No. 1, February, 1959, pp. 47-60.

7. Some notion of how migrant experiences have influenced the development of a time perspective for those Mexican-Americans and Negroes who have been in the migrant labor cycle may be obtained from Dorothy Nelkin's "Unpredictability and Life Style in a Migrant Labor Camp," Social Problems, Vol. 17, No. 4, Spring, 1970, pp. 472-487. As she puts it on p. 480, "Time is not perceived as a continuous and predictable process, but as a series of disconnected periods; of good seasons and bad seasons, good weeks and bad weeks. What happens during the current week or season is not perceived as having much to do with what will happen during a future period." In sum, the sense of causality, the relation between effort and return, is perceived in terms of an environment which may be neither predicted nor controlled. The entire article is extremely helpful in understanding how the migrant experience may have a profound influence on the world view of some of our respondents, particularly those Mexican-Americans who have been in the agricultural migrant labor cycle. In referring to the manner in which leisure time is spent, Nelkin points out, on page 485, that "Story telling in migrant camps, as a means of entertainment requiring little equipment or energy, occupies a great deal of leisure time . . they [stories] reflect the ambiguity and lack of predictability in the migrant labor system. For example, a thread of luck or fate runs through most tales; and, in a content analysis of 148 stories collected in migrant labor camps, 24 percent of the themes had to do with getting away with something or 'beating the system' in some indirect way. Again, since one cannot control or confront the system, it must be defied or circumvented."

8. In a general sense, the world view of the inmigrant Negro can be understood in the same way that Gottlieb and Ramsey presented it in *Understanding Children of Poverty,* Chicago: Science Research Associates, 1967. They have reprinted, commencing on page 20, a section entitled "Work Patterns of the Deprived," reprinted from ETC: A Review of General Semantics, Vol. III, No. 4, 1946. This particular excerpt is entitled "Clark Gives Up Trying," and is found on page 27: "For the worker, short periods of good wages and plentiful jobs do not take the place of this security. One cannot change his way of living or [prompt him to] buy a home or educate his children, on this kind of income. To have a chance to develop stable habits of living, which means good work habits, people must have a stable job. The underprivileged worker is perfectly realistic when he asks, 'Why should I try to save and get a little ahead in these times, when I'll be back on relief, anyhow, in a year or two?' "

For the reader who believes that the past should be forgotten and that the urban Negro should have an over-all view based strictly on what the white man now tells him the world is like, perusal of *Lay My Burden Down: A Folk History of Slavery,* E. A. Botkin (ed.), Chicago: University of Chicago Press, 1945, should be helpful. In Part 2, "Long Remembrance," p. 59, Botkin states, "The narrators testify repeatedly to the freshness, vividness, and concreteness of their memories. 'Does I 'member much 'bout slavery times? Well, there is no way for me to disremember unless I die.' 'I got 'membrance like they don't have nowadays. That 'cause things is going round and round too fast without no setting and talking things over.' 'I 'members now clear as yesterday things I forgot for a long time.' 'I remember that day just as good as it had been this day right here.' 'I recollects just as bright as the stars be shining.' "

"When there are fewer things to remember, as is the case in a restricted world, one remembers things more vividly, especially little things. And it is the little things—associations of events with objects, images, and sensations—that, in the absence of written records, help one to remember."

This volume is one of the best that may be found for those who wish to obtain some idea of how the Negro's past influences his present view of the world.

9. Madsen, op. cit., p. 17. The importance of the family is described by Madsen as follows: "The most important role of the individual is his familial role and the family is the most valued institution in Mexican-American society. The individual owes his primary loyalties to the family, which is also the source of most affective relations. Gregorio said, 'I owe everything to my family. Were it not for my parents' love for each other, I would never have been born. They raised me and taught me all I know. They have protected me and in my parents' home I know I will always find love and understanding. When one has a family, one is never alone nor without help in time of need. God created the family and one way to show respect to Him is to respect one's parents.' The worst sin a Latin can conceive is to violate his obligations to his parents and siblings." Also see Kluckhohn and Strodtbeck, op. cit., pp. 17-20.

10. The validity of the world view scale in differentiating respondents was described in detail in the following paper: Lyle W. Shannon, "The Study of Migrants as Members of Social Systems," Proceedings of the Annual Spring Meeting of the American Ethnological Society, 1968, University of Washington Press, Seattle, pp. 34-64.

11. Certainty of responses was measured on a four-point scale: (1) understood questions—response well formulated; (2) understood questions—response not well formulated; (3) predominantly acquiescive response pattern; and (4) confused, failed to answer question. Claire Peterson (Vanderbilt University) selected the subsample and constructed tables that served as a starting point for the analysis.

12. A series of tests of the significance of differences between samples and subgroups within samples is presented below. Since three separate samples were independently selected, it is also appropriate to test for the significance of differences between samples, two at a time. Generally speaking, unless a difference is significant at the .05 level the difference is not considered significant, although χ^2 values indicating a lower level of confidence to .10 are reported.

Tests of Significance of Differences Between World View in S1 and S2 and Subgroups of S1 and S2

Mexican-American S1 - S2	$\chi^2 = .11$, 1 d.f., not significant
Negro S1 - S2	$\chi^2 = .68$, 1 d.f., not significant
Anglo S1 - S2	$\chi^2 = .60$, 1 d.f., not significant
Mexican-American/Negro/Anglo S1	$\chi^2 = 102.18$, 2 d.f., p < .001
Mexican-American/Negro/Anglo S2	$\chi^2 = 76.78$, 2 d.f., p < .001
Mexican-American/Anglo S1	$\chi^2 = 91.74$, 1 d.f., p < .001
Mexican-American/Anglo S2	$\chi^2 = 67.94$, 1 d.f., p < .001

Negro/Anglo S1 $\chi^2 = 65.14$, 1 d.f., $p < .001$
Negro/Anglo S2 $\chi^2 = 42.42$, 1 d.f., $p < .001$

Mexican-American/Negro S1 $\chi^2 = 2.38$, 1 d.f., not significant
Mexican-American/Negro S2 $\chi^2 = 2.59$, 1 d.f., not significant

13. Tests of Significance of Differences Between Level of Aspiration for Children in S1 and S2 and Subgroups of S1 and S2

Mexican-American S1 - S2 $\chi^2 = .52$, 4 d.f., not significant
Negro S1 - S2 $\chi^2 = 1.31$, 4 d.f., not significant
Anglo S1 - S2 $\chi^2 = .07$, 4 d.f., not significant

14. Tests of Significance of Differences Between Vocational Aspirations for Children in S1 and S2 and Subgroups of S1 and S2

Mexican-American S1 - S2 $\chi^2 = 1.39$, 4 d.f., not significant
Negro S1 - S2 $\chi^2 = .75$, 4 d.f., not significant
Anglo S1 - S2 $\chi^2 = .68$, 4 d.f., not significant

15. Years of Education for Respondents, by Percent

| | Mexican-American | | Negro | | Anglo | |
	S1	S2	S1	S2	S1	S2
Less than 9	76	78	46	39	22	14
9 or more	24	22	54	61	78	86
	100	100	100	100	100	100

Tests of Significance of Difference Between Years of Education in S1 and S2 and Subgroups of S1 and S2

Mexican-American S1 - S2 $\chi^2 = .12$, 1 d.f., not significant
Negro S1 - S2 $\chi^2 = 1.51$, 1 d.f., not significant
Anglo S1 - S2 $\chi^2 = 4.96$, 1 d.f., $p < .05$

16. Tests of Significance of Differences Between Total Annual Family Income in S1 and S2 and Subgroups of S1 and S2

Mexican-American S1 - S2 $\chi^2 = .11$, 2 d.f., not significant
Negro S1 - S2 $\chi^2 = 1.49$, 2 d.f., not significant
Anglo S1 - S2 $\chi^2 = .95$, 2 d.f., not significant

17. Tests of Significance of Differences Between Length of Residence in Racine in S1 and S2 and Subgroups of S1 and S2

Mexican-American S1 - S2 $\chi^2 = 1.45$, 1 d.f., not significant
Negro S1 - S2 $\chi^2 = .02$, 1 d.f., not significant
Anglo S1 - S2 $\chi^2 = 2.37$, 1 d.f., not significant

18. Having selected a world view sample, one for which there could be some assurance that responses to the world view questions were judged to be meaningful to respondents, the relationship of world view to a number of other variables such as time in the community, education, and income was carefully examined. See Lyle W. Shannon, "The Study of Migrants as Members of Social Systems," op. cit., pp. 34-64.

We hypothesized that an active world view was associated with long residence, high education, and high income, and a passive world view with the opposite. The rank ordering of Mexican-Americans, Negroes, and Anglos on world view, education, and income, and to some extent, length of residence should be related to world view in a systematic fashion.

The Relation of World View to Education and Length of Residence

			Individual- istic, Active	Group- Oriented, Passive	Total
MEXICAN-AMERICAN					
Low Education	SR	N=113	34	66	100
	LR		32	68	100
High Education	SR	N=32	80	20	100
	LR		50	50	100
NEGRO					
Low Education	SR	N=62	38	62	100
	LR		41	59	100
High Education	SR	N=95	50	50	100
	LR		57	43	100
ANGLO					
Low Education	SR		0	100	100
	LR	N=29	62	39	101
	All		69	31	100
High Education	SR		85	15	100
	LR	N=185	87	13	100
	All		83	18	101

SR: Short residence; less than 10 years; LR: Long residence, 10 years or more; All: always lived in Racine; Low education: less than nine years; High education: nine years or more.

The data shown in the preceding table may be used to generate two different series of statistics. One series deals with the relationship of time in the community or of length of residence to world view and the other series deals with the relationship of education to world view; in each case the influence of one variable on world view may be observed with the other controlled. With race or ethnicity controlled, in no case was there a significant relationship between world view and length of residence in the community. And with controls introduced for race or ethnicity and years of education, world view and length of residence produced no statistically significant relationships. The largest correlation, a modest one and in the opposite direction of the general hypothesis that persons longest in the community would have the most active world views, was .28 for Mexican-Americans (usually categorized as having "passive" world views) in the 10 or more years of education category, but was not statistically significant. On the other hand, this correlation made sense, considering the characteristics of the Mexican-American sample retrospectively; younger, better educated, new arrivals were actively seeking economic opportunities. Although age of respondent did have some influence on answers to questions, this varied from one area of concern to another. For example, with age of male controlled, occupational level varied significantly with race or ethnicity in the larger sample. Similarly, with age of male controlled, income varied significantly by race or ethnicity. Age of respondent and age of male had scattered correlations with other variables, but there were neither uniformities nor patterns that would lead to the conclusion that age should be controlled in the analysis at hand.

When world view scale scores and years of education were correlated for the combined samples without length of residence controlled, there was a statistically significant but modest correlation of .34 between education and world view; with time in the community controlled, education continues to have a significant relationship to world view. When controls were introduced for race or ethnicity, years of education and world view are significantly related among the Mexican-Americans and Anglos but not among the Negroes. When race or ethnicity and length of residence were controlled, seven tables could be produced; the relationship between world view scale scores and education remained in every table but was statistically significant in only two cases, among those Anglos and Mexican-Americans who had lived in Racine nine years or less. On a strictly statistical basis one must conclude that respondents with a better education were neither more individualistic nor active than those with less education—but, considering the general pattern of correlations and the two significant correlations among short term residents, education was more closely related to world view than was time in the community. Short residence, better educated Mexican-Americans had more active world views than did their short-residence, less educated Mexican-American counterparts. The same relationship also existed for short-residence Anglos and, although not statistically significant, was apparent to a lesser degree among the Negroes. But one only need examine the place of origin of most Negro versus most Mexican-American and Anglo respondents in order to gain some idea of how the influence of years of formal education on world view could well have been washed out by other variables in their environment prior to recent arrival in Racine.

Although the hypothesis that length of residence in the community and education are closely related to world view must be rejected, particularly length of residence, it was quite apparent that both, in concert with race or ethnicity, did have some relationship to world view but that length of residence played a minor part in it. The statistically significant relationship between world view and education disappeared in most cases when controls were introduced, because the uncontrolled correlations were generated by the fact that Anglos tended to have active world view scores, be well educated, and be long time residents of the community, while the Negroes and Mexican-Americans had the least education, the shortest period of residence, and the most passive world views. In sum, the greatest contrast in world view was between passive, long residence, least educated Mexican-Americans and active, long residence, most educated Anglos.

Considering the distribution of income and world view on race or ethnicity, as shown below, it was not surprising that, when the samples were examined before controlling for ethnicity, the number of high income persons with active scores was over twice those with passive scores, while twice as many low income persons had passive as had active scores. When low, medium, and high income persons were compared, holding ethnicity constant, world view varied significantly with income within the Mexican-American and Anglo samples but not within the Negro sample. In each case, low income persons were more likely to be passive and high income persons more likely to be active. But when two of the three income groups within each race or ethnic group were compared, only the high and low income groups were significantly different in world view in the Mexican-American group, the low and high income groups did not reach the .05 level among the Negroes, and only two pairs were significantly different in the Anglo group.

However evident it was that the world view scale differentiated between Mexican-Americans, Negroes, and Anglos, the distribution of scores was so skewed toward the lower end of the scale for Mexican-Americans in particular, that subcategories of Mexican-Americans could scarcely be differentiated by their world view scores.

When income was controlled, the world view of Mexican-Americans, Negroes, and Anglos continued to differ significantly, with world view having its closest relationship to ethnicity within the high income group. When race or ethnic groups were compared with income controlled, the Mexican-American and Anglo groups differed significantly on world view within each income category, but the Mexican-American and Negro groups never differed significantly. Just as stated in the previous paragraphs, differences in world view between ethnic groups within income groups were somewhat greater than the differences in world view between income groups within ethnic groups. Although high income persons tended to be active and low income persons tended to be passive,

The Relation of World View to Income, by Percent

	Mexican-American			Negro			Anglo		
	L	M	H	L	M	H	L	M	H
Individualistic, Active	31	40	57	40	45	59	59	71	87
Group-Oriented, Passive	69	60	44	61	55	41	41	29	13
	100	100	101	101	100	100	100	100	100
N=	81	30	46	43	29	75	22	31	143

Low (L) = less than $4,499; Medium (M) = $4,500 - $5,499; High (H) = $5,500 or more.

variation based on ethnicity within income groups was more marked than variation based on income within race or ethnic groups.

19. For examples of earlier writings on the relationship of religion to other variables, see: Max Weber, *The Protestant Ethic and the Spirit of Capitalism,* translated by Talcott Parsons, New York: Charles Scribner's Sons, 1930; Werner Sombart, *The Jews and Modern Capitalism,* translated by M. Epstein, London: F. Unwin, 1913; R. H. Tawney, *Religion and the Rise of Capitalism,* New York: Harcourt Brace, 1926. More recently the relationship of religious identification to attitudes and behavior pertinent to the present discussion have been dealt with in the following: Gerhard Lenski, *The Religious Factor,* New York: Doubleday, 1961; Raymond W. Mack, Raymond J. Murphy, and Seymour Yellin, "The Protestant Ethic, Level of Aspiration, and Social Mobility: An Empirical Test," American Sociological Review, 21(June, 1956), pp. 295-300; Albert J. Mayer and Harry Sharp, "Religious Preference and Worldly Success," American Sociological Review, 27 (April, 1962), pp. 218-227.

20. For one study examining the relationship of religion to militancy among Negroes, see Gary T. Marx, *Protest and Prejudice: A Study of Belief in the Black Community,* New York: Harper & Row, 1969, Chapter 4, "Religion: Opiate or Inspiration of Civil Rights Militancy?" pp. 94-105. Militancy varied from 43 percent of the Episcopalians, 36 percent of the Catholics, 28 percent of the Methodists, 25 percent of the Baptists, to 15 percent of the Sects and Cults. Militancy was least among those who considered religion extremely important (22 percent) and greatest (62 percent) among those who considered it not at all important. Militancy was least among those who attended church more than once a week (18 percent) and greatest among those who attended less than once a year (32 percent). When these and various items were combined into an index of religiosity, only 19 percent of the very religious were militant and 49 percent of the not at all religious were militant. The author concluded by stating that "Until such time as religion loosens its hold over these people, or comes to embody to a greater extent the belief that man as well as God can bring about secular change, and focuses more on the here and now, religion would seem to be an important factor working against the widespread radicalization of the Negro public."

21. For a detailed comparison of every possible combination of race and ethnicity, religion, and sex, see Lyle W. Shannon, "The Economic Absorption and Cultural Integration of Inmigrant Workers: Characteristics of the Individual vs. the Nature of the System," Chapter 11 in *Migration and Human Adaptation,* Eugene B. Brody (ed.), Beverly Hills: Sage Publications, 1970.

22. If expressed in terms of an r_4 coefficient of correlation, these would be close to or above .7000.

23. As James N. Morgan puts it in "The Achievement Motive and Economic Behavior," Economic Development and Cultural Change, Vol. XII, No. 3, April, 1964, pp. 243-267, "Why should achievement motivation affect the hourly wages of an individual who because of his education or race or sex has open to him only jobs where the pay scale is fixed, perhaps by union contract?" The notion that motivation to achieve is the independent variable has been attractive to those who would prefer not to look at how the system works for the less fortunate in our society, that is, less fortunate in terms of the fact that they have

ascribed statuses that handicap them in their efforts to become absorbed into the economy. Morgan further states that ". . . in general the results suggest that the achievement motive and beliefs about the probabilities of hard work being rewarded are related to the economic behavior of individuals within a culture."

24. There are, of course, various interpretations of Catholicism. Madsen, op. cit., p. 58, states, "The majority of Mexican-Americans are Catholics but their interpretations of Catholicism vary with class and education. The Catholicism of the conservative elite is orthodox and sophisticated. The lower class and a large part of the middle class hold to Spanish-Indian beliefs derived from Mexican folk Catholicism. Such beliefs are the despair of the priests. Sermons delivered in the Magic Valley urge the Mexican-Americans to abandon their 'superstitions' and accept church dogma. The Mexican-Americans listen politely to these sermons and ignore them."

25. In a survey of the perceptions and attitudes of more than 5,000 Negroes and whites in 15 major American cities, the question was asked, "If a young Negro works hard enough, do you think he or she can usually get ahead in this country in spite of prejudice and discrimination, or that he doesn't have much chance no matter how hard he works?" Seventy-eight percent of the Negroes were in agreement with this statement. When Campbell and Schuman controlled for age and education, they found that education has a clear positive association with the belief that a Negro can get ahead despite prejudice and discrimination. The relationship was greater for young men in the twenties and thirties where belief in individual accomplishment was held by 93 percent of the college graduates but only 68 percent of those with a grade school education. According to the authors, a merging of age and education trends could mean that Negro males believe in the value of individual initiative and in the possibility of individual achievement and the reinforcement of these beliefs in those who manage to go to school. As they stated, "The more he gets ahead, the more he thinks he should be able to get ahead. But what is often called the school drop-out lacks the possibility of achievement, and apparently in a growing proportion of cases he believes that it is the society that is at fault, not he himself." Angus Campbell and Howard Schuman, *Racial Attitudes in Fifteen American Cities,* a Preliminary Report Prepared for the National Advisory Commission on Civil Disorders, Ann Arbor: Institute for Social Research, 1968, pp. 27-28.

26. While we refer in an academic and theoretical manner to such systemic forces as the occupational or educational opportunity structures of society, Mexican-Americans and Negroes may have more concrete forces and events in their minds. Stories of threats of death and death at the hands of the Texas Rangers are still told by the Mexican-American residents in the lower Rio Grande Valley area. The influence of such events on the development of world view is illustrated by Arthur J. Rubel in Chapter 2, "New Lots in Historical Context," from his volume, *Across the Tracks: Mexican-Americans in a Texas City,* Austin: University of Texas Press, 1966, pp. 25-54.

27. Of the numerous volumes of readings that have been published during the last few years and which treat the culture of poverty among other topics, one of the most useful is that edited by Nona Y. Glazer and Carol F. Creedon, *Children and Poverty: Some Sociological and Psychological Perspectives,*

Chicago: Rand McNally & Co., 1968. While the entire volume is excellent, several articles are particularly pertinent in sensitizing the reader to the world view that Negro inmigrants have acquired because of their life experiences prior to arriving in Racine. Particularly helpful is Mary Frances Greene and Orletta Ryan's "The Schoolchildren: Growing Up in the Slums," pp. 73-77; "Cries of Harlem," HARYOU, pp. 78-85; and "Mississippi Black Paper," by Misseduc, pp. 86-94. Each of these gives us some idea of how, through the process of socialization in either Mississippi or the black ghetto, the Negro acquires a view of how the society is organized and what his chances are in relation to the chances of his more fortunate white competitors.

Chapter 10

ASSOCIATES AND ASSOCIATIONS

The associational patterns of the Mexican-American, Negro, and Anglo respondents were hypothesized to be important differentiating variables in the urban, industrial milieu. We assumed that the bonds tying the inmigrant to his former way of life would be so pervasive in nature that they would exert a powerful influence on ethnic values and behavior in a northern industrial setting regardless of the time sequence or of the indirect or personal form of contact.

Association with Relatives and Close Friends

In 1959, over half the relatives of Mexican-Americans but only one-third of the Anglos lived outside of the Midwest. Of the Mexican-Americans, most stated that their families lived in Texas. Well over half the Anglos in the 1960 sample reported that most members of the family lived within the boundaries of the city of Racine. Only one-fifth of the Mexican-Americans and one-third of the Negroes gave similar responses. Relatives of the Mexican-Americans were still in Texas and more of the relatives of the Negroes were in Mississippi than in any other state.

A scale measuring ties to former home (Table 48) demonstrated that respondents from the 1960 sample with the closest ties to their former home had the following characteristics: (1) kept in touch with relatives not living in Racine; (2) did not plan to go back to former home to live;[1] (3) visited former home more than once a year, or at the minimum every two years; and (4) were less satisfied

TABLE 48. TIES TO FORMER HOME, 1960 (IN PERCENTAGES)

	Scale Type	Mexican-American	Negro	Anglo
Minimum ties	0	11	2	5
	1	21	8	7
	2	51	38	32
	3	11	41	45
Maximum ties	4	6	11	10
		100	100	99

R = .9292; MR = .7550; Mexican-American/Anglo χ^2 = 24.28, 1 d.f., p <.001; Negro/Anglo χ^2 = .37, 1 d.f., not significant; Mexican-American/Negro χ^2 = 24.24, 1 d.f., p <.001. Sufficient data were available to generate scale scores for only 438 persons.

with Racine than with former home—with or without reservations. Surprisingly, Negroes and Anglos had significantly closer ties to their former home than Mexican-Americans.[2]

Since the mere presence of relatives either visiting or living in a community tells us little about the degree to which one is tied to them or the potential influence of these ties, respondents were asked to indicate the number of relatives visited currently, their relationship to relatives who were visited often, and the occupational status of these relatives. There was no meaningful diversity between the samples with respect to the number of relatives visited.[3] The only instance of statistical significance in the relationship of relatives who were visited often was in 1961 when Anglos (36 percent) were more likely than Negroes to visit their siblings.

An examination of the occupational status of these relatives identified the most striking dissimilarity between samples. Anglo relatives were significantly higher in occupational status than those of the Mexican-Americans in 1959, higher than the Mexican-Americans and Negroes in 1960, and higher than the Negroes in 1961. What resulted, of course, was a visiting pattern based on differences in the occupational distribution of each ethnic group. Since the pattern emphasized socializing within one's occupational level, there was little likelihood that it would generate an exchange of information leading to occupational mobility among the Negroes and Mexican-Americans. The findings were in the same direction regardless of where the relatives lived—within or outside the community.[4]

In 1960 each respondent was asked about the friends he saw often in Racine. The occupational level pattern for first and second-mentioned friends was similar to that for relatives. Anglos and their

TABLE 49. OCCUPATIONAL STATUS OF ASSOCIATES SCALE (IN PERCENTAGES)

	Scale Type	1960			1961	
		Mexican-American	Negro	Anglo	Negro	Anglo
Low occupational status	0	77	66	25	29	11
	1	11	12	8	51	11
	2	8	13	14	9	14
	3	3	3	11	9	20
	4	0	3	18	1	17
High occupational status	5	1	3	24	0	28
		100	100	100	99	101

1960: R = .8903; MR = .7500; Mexican-American/Anglo χ^2 = 60.46, 1 d.f., p < .001; Negro/Anglo χ^2 = 78.69, 1 d.f., p < .001; Mexican-American/Negro χ^2 = 2.79, not significant.

1961: R = .9012; MR = .6640; Negro/Anglo χ^2 = 109.42, 1 d.f., p < .001.

Sufficient data were available to generate scale scores for only 558 persons in 1960.

friends were working not only at similar occupational levels but at levels higher than either Mexican-Americans or Negroes and their friends. Five out of six friends of the Mexican-Americans had Spanish surnames.[5]

Occupational status of associates scales were devised from the 1960 and 1961 responses identifying friends (first mentioned) in Racine, relatives (first mentioned) in and outside Racine, and relatives (second mentioned) outside Racine. In either year, Anglos had associates of higher occupational status than Mexican-Americans or Negroes; the difference between Negroes and Mexican-Americans was not significant (Table 49).

Anglo/Negro Association

A separate set of questions on associative behavior for Negroes in 1960 is presented in Table 50. Half the Negroes in Racine stated that they had non-Negro friends but only 16 percent reported that the proportion of non-Negro to Negro friends was one-half or more. Over half the Negro respondents invited non-Negroes to visit their homes for predominantly social reasons.[6] One of the questions was, "Do you think that Negroes in Racine should have more contacts with non-Negroes, less, or about the same?" Sixty-five percent of the Negroes thought that there should be more contact between Negroes and non-Negroes, 23 percent wanted the same amount of contact as before, and only 9 percent thought that there should be less contact.[7]

TABLE 50. CHARACTERISTICS OF ANGLO/NEGRO ASSOCIATION AS DESCRIBED BY NEGROES, 1960 (IN PERCENTAGES)

	Male	Female
"Among your friends in Racine, are there any who are not Negroes?"		
Yes	55	46
No	45	55
	100	101
"About what proportion of your friends are not Negro?"		
Most not Negro	4	6
About half and half	13	9
Mostly Negro	32	27
All Negro but one	5	5
All Negro friends	45	54
	99	101
"Have you ever invited persons who are not Negroes to visit your home?"		
Yes	57	54
No	43	46
	100	100
"Have you ever invited persons who are not Negroes to visit your home—on what occasions?"		
Informal social occasions	34	26
Business occasions	3	1
Contacts through children	3	4
Religious contacts	5	4
No particular occasion	12	16
Other	0	2
Not ascertained or does not have non-Negro friends	43	46
	100	99
"Do you think that Negroes in Racine should have more contacts with non-Negroes, less, or about the same?"		
More	74	58
Less	9	10
Same	14	30
Not ascertained	4	2
	101	100

In 1961 the wording of the last question was changed somewhat: "Would you personally like to have more social contact with non-Negroes, less, or about the same?" The data in Table 51 indicate an almost complete reversal of opinion from that in 1960. The majority of the respondents (66 percent) wanted to keep the same degree of social contacts with non-Negroes as before and only 28

TABLE 51. NEGRO ATTITUDES TOWARD INCREASED SOCIAL CONTACT BETWEEN NEGROES AND NON-NEGROES, 1961 (IN PERCENTAGES)

	Male	Female
"Would you personally like to have more social contact with non-Negroes, less, or about the same?"		
More	41	18
Less	5	4
Same	53	76
Don't know, not ascertained	2	3
	101	101

Negro Female/Negro Male χ^2 = 7.44, 1 d.f., p < .01.

percent wanted more. This surprising difference might be attributed to the change in wording of the question with emphasis on "more *social* contact," and the tone of the question, which implied a personal rather than an impersonal reaction. Negro males desired contact with non-Negroes in twice as many cases as did Negro females. Since Negro females were oriented toward their neighborhoods to a greater extent than the males, they might have had more satisfying relationships in the Negro community than the males and had less reason to extend their contacts to whites. Or, the greater mobility orientation of Negro males as compared to females might permit them to have Negro-white contacts more easily as a means of attaining social or economic advancement.

When the respondents were asked to give reasons, those desiring more contact usually mentioned such factors as the promotion of better understanding; those desiring less contact mentioned a mutual dislike between races or the advisability of staying in one's place; and those desiring the same amount of contact reported general satisfaction with the "status quo."

Gary T. Marx has more recently pointed out in *Protest and Prejudice: A Study of Belief in the Black Community* that although many studies have shown that as age decreases among white Americans, tolerance increases, this has not been the case for his data. The exact opposite relationship was found—the younger the Negro, the more anti-white his attitude. For example, 27 percent of those under 30 had anti-white attitudes, decreasing to 19 percent among those between the ages of 60 and 75, and only 14 percent for those 75 and above.[8] Marx also points out that persons with higher indexes of social participation had a lower degree of hostility towards whites. The variation ranged from 13 percent not being anti-white among those with the lowest index of social participation to 36 percent not being anti-white among those with the highest index of social participation. Although index of social participation is related to education and position in the class structure, social participation has an effect on anti-white attitudes independent of social class.[9] Marx speaks of anti-white and anti-black hostility as being related to less sophistication. The more sophisticated the Negro, the less his anti-white attitude. On a six-point scale of intellectual sophistication, Marx found variation from 9 to 36 percent in anti-white attitudes. On a four-point scale of acceptance of intellectual values, anti-white attitudes ranged from 15 percent to 32 percent.[10]

A Negro male respondent in Racine who wanted more contact said:

"We could get to know more about the people you see. It would give more understanding between people. All this trouble comes from no understanding."

A Negro female who wanted less contact stated:

"Who wants to socialize with someone who doesn't want to socialize with you? I don't force myself on my own people. They don't get along. The two peoples (races) have different ideas. A slave can't get along with his master. One has to be over the other. It always has been like that and always will be."

One wonders if the above respondent would speak in this manner today. Another Negro male felt contacts should remain about the same because:

"I think we have about enough. Sometimes too much contact goes too far. Like if you go to too many parties together you know what can happen."

In 1961, an additional question asked whether the Negroes would like their children to have more contact with non-Negro children (Table 52). Almost one-third of the respondents desired more contact while close to two-thirds of the respondents favored the same degree of contact as before. Reasons for advocating more, less, or the same amount of interaction with non-Negroes were essentially the same as those stated in response to other questions on contact.

A scale measuring the extent of community integration among Negroes and Anglos (Table 53) was designed in 1961. The respondent with the highest score had relatives in Racine, was able to verbalize about the relationship of relatives whom he frequently saw,

TABLE 52. NEGRO ATTITUDES TOWARD INCREASED SOCIAL CONTACT BETWEEN CHILDREN AND NON-NEGRO CHILDREN, 1961 (IN PERCENTAGES)

	Male	Female
"Would you like your children to have more contact with non-Negro children, less, or about the same?"		
More	39	22
Less	2	3
Same	56	71
Don't know, not ascertained	3	5
	100	101

Negro Female/Negro Male χ^2 = 3.80, 1 d.f., not significant.

had especially good friends in Racine, and considered his neighbors as very close friends. The pattern of responses indicated that Negroes were significantly more integrated into the local community (Racine) than Anglos. This may be as has been suggested, because discriminatory conditions outside a Negro community create a degree of solidarity within the Negro ghetto that is not commonly found among the larger population. In addition, the Negroes in Racine (as elsewhere) resided for the most part in an overcrowded environment which encourages primary group type relationships.

Persons Sought for Advice

In 1959, questions were designed to draw forth information indicating the characteristics of associates whose advice was sought by Mexican-American and Anglo respondents. When the responses were coded (Table 54) according to the relationship of the first-mentioned person sought for advice, there was no significant difference between Mexican-Americans and Anglos. Husbands were

TABLE 53. COMMUNITY INTEGRATION SCALE, 1961 (IN PERCENTAGES)

	Scale Type	Negro	Anglo
Minimum integration	0	3	6
	1	5	4
	2	7	14
	3	23	31
Maximum integration	4	62	44
		100	99

R = .9279; MR = .7450; Negro/Anglo χ^2 = 9.74, 1 d.f., p < .01.

TABLE 54. CHARACTERISTICS OF PEOPLE WHOSE ADVICE IS RELIED UPON, 1959 (IN PERCENTAGES)

"Of all the people that you know, whom do you go to when you want to talk things over? I mean whose advice would you *most* rely on?"

Relation

	Mexican-American		Anglo	
	M	F	M	F
Relative	43	38	29	44
Acquaintance	2	15	12	15
Professional person; (doctor, priest, lawyer)	9	11	6	5
Wife, husband	24	16	37	28
Combination, other	1	2	0	0
Not ascertained	21	17	15	8
	99	100	99	100

Mexican-American/Anglo χ^2 = 1.59, 1 d.f., not significant.
Mexican-American Male/Mexican-American Female χ^2 = 12.00, 3 d.f., p < .01; Anglo Male/Anglo Female χ^2 = 4.35, 3 d.f., not significant; Mexican-American Male/Anglo Male χ^2 = 11.83, 3 d.f., p < .01; Mexican-American Female/Anglo Female χ^2 = 5.50, 3 d.f., not significant.

Occupation

	Mexican-American		Anglo	
	M	F	M	F
Professional, technical, managerial, proprietor, clerical, sales	12	15	24	32
Craftsman, foreman	8	3	10	13
Operative, industrial labor, maintenance and service	11	17	10	14
Housewife	27	24	33	23
Agricultural laborer	2	1	4	0
Not ascertained	41	40	19	19
	101	100	100	101

Mexican-American/Anglo χ^2 = 8.63, 1 d.f., p < .01.

more likely than wives, and Anglos were more likely than Mexican-Americans, to report that they turned to their spouses for advice. But when responses were coded for occupational level of first mentioned person the two groups differed significantly with Anglos seeking out people of higher occupational status.

Both Mexican-Americans and Anglos turned for advice to high occupational level persons more often than would be expected in relationship to the number of persons in high occupational status listed as their close associates (friends or relatives) either inside or outside Racine. In other words, the Mexican-Americans and Anglos selected upward (more so) when choosing the people from whom they wanted advice than when seeking associates in day-to-day living.

More specific questions concerning "advisers" were asked of each sex. To whom did males talk about problems concerning work? Both Anglo and Mexican-American males were likely to talk to their wives, relatives, or supervisory personnel on the job; within these categories, Anglos were more likely to talk to their wives and Mexican-Americans were more likely to converse with union officials (Table 55). We should add that Mexican-Americans sought advice from persons with Mexican rather than non-Mexican surnames at a ratio of approximately three to one. When Mexican-American females were asked who advised them on the health of their children, the ratio of persons with Mexican to those with non-Mexican surnames was 25 to one. When asked who advised them before they made an expensive

TABLE 55. PEOPLE SOUGHT FOR ADVICE IN WORK SITUATIONS, MALES ONLY, 1959 (IN PERCENTAGES)

	Relationship	
	Mexican-American Males	Anglo Males
"If a problem comes up concerning *work*, whom do you talk to about it?"		
Wife	10	19
Parent	5	1
Other relative	14	10
Union official	10	5
Supervisor, foreman, or other	17	23
Co-worker	2	9
Employment office	4	0
Friends who are not co-workers	2	3
Other	2	0
Not ascertained, inapplicable	35	29
	101	99

purchase, the ratio of Mexican to non-Mexican surnames was five to one. It seemed very evident to us that Mexican-Americans were taking advice from other Mexican-Americans rather than from urban, industrial Anglo models. The willingness of Mexican-Americans to turn to other Mexican-Americans for advice in various problem situations should not be considered as evidence of the existence of well-developed leadership and a multiplicity of formal or voluntary organizations in the Mexican-American community.[11] There were very few Mexican-American organizations at the time of the 1960 interviews, although we shall comment later on some changes that have occurred in the Mexican-American community in Racine since that time, as well as pertinent changes that have taken place on the national scene.

Voluntary Associations and other Leisure Time Activities

Another facet of associative behavior, but one which is never easy to measure, is the use of leisure time and its relationship to contacts with the larger society. In each year of the survey we employed a series of questions about such leisure time activities as attendance at dances or parties, club membership, and sports and hobbies. Each respondent was asked to describe participation of husband and wife in these activities. We expected that the 1959 Anglo sample, which was skewed toward the lower end of the social class, would show a low rate of social participation similar to that hypothesized for the Mexican-Americans.[12]

In all cases of significant differences in patterns of social participation between the samples during the three years of the study, attendance at dances or parties was the greatest, with Anglos reporting greater attendance (56 percent) than did Mexican-Americans (46 percent) or Negroes (40 percent). We agreed that the disparity was not entirely because of ethnic or racial differences but more because of the socioeconomic, marital, or family status of respondents.[13]

All groups attended movies in approximately the same proportion, but a greater number of Mexican-Americans (61 percent) reported being present at movies than either Negroes (59 percent) or Anglos (53 percent). Significant differences between the samples appeared in 1959 when more Mexican-American than Anglo husbands reported movie attendance and in 1960 when more Mexican-American than Negro husbands went to them.

A question on television viewing was included in 1961. Negro husbands reported significantly longer hours watching television than Anglo husbands. The same was true in comparing the wives, but both Negro and Anglo wives reported even longer hours watching television than did males.

In 1959, 8 percent of the Mexican-Americans and 39 percent of the Anglos belonged to clubs, a statistically significant difference. In one of the surveys conducted by the Detroit Area Study of a cross-section of the Detroit population, it was shown that nearly two-thirds of the population were members of formal groups. While only 50 percent belonged to only one group, 26 percent of the total belonged to two, 13 percent belonged to three, 6 percent to four, 3 percent to five, and 2 percent to six or more. Church affiliation was not defined as a membership in a formal group in this study. It should, of course, be noted that of the 37 percent who did not belong to a formal group 17 percent did belong to a church, so it may be said that only 20 percent of the total were completely non-members.[14] In 1960, 14 percent of the Mexican-Americans, 17 percent of the Negroes, and 49 percent of the Anglos reported membership in clubs; the difference between Anglos and Negroes or Mexican-Americans was statistically significant. A similar pattern was obtained when respondents were questioned about club attendance.[15] Anglo presence in clubs was higher than Mexican-American in 1959 and 1960, and higher than Negroes in 1960 and 1961.

Respondents were queried about participation in church organizations during each year of the survey. In 1959 and 1960, Anglo husbands and wives participated in church organizations to a significantly greater extent than either Negro or Mexican-American husbands and wives.[16] While there was no significant difference in interest shown by Mexican-American (8 percent) or Negro (11 percent) husbands, Negro wives (26 percent) participated in church organizations to a significantly greater extent than Mexican-American (10 percent) wives.[17] In 1961, Anglo wives (35 percent) and husbands (32 percent) participated to a significantly greater extent than Negro wives (5 percent) and husbands (7 percent).

The only leisure time activity which sharply discriminated between the sexes was attendance at public drinking houses (Table 56). The sizeable difference between Mexican-American males and females who frequented taverns (statistically significant) was not

TABLE 56. ATTENDANCE AT PUBLIC DRINKING HOUSES (IN PERCENTAGES)

| | 1959 | | 1960 | | | 1961 | |
	Mexican-American	Anglo	Mexican-American	Negro	Anglo	Negro	Anglo
"Do you ever go to taverns?"							
MALES							
Yes	56	43	55	48	54	52	51
No	34	39	33	37	40	35	45
Inapplicable	10	19	11	15	6	13	4
	100	101	99	100	100	100	100
FEMALES							
Yes	10	30	16	33	45	34	46
No	86	65	84	65	55	64	53
Inapplicable	5	5	1	2	1	2	1
	101	100	101	100	101	100	100

MALES:
1959: Mexican-American/Anglo χ^2 = 2.64, 1 d.f., not significant.
1960: Mexican-American/Anglo χ^2 = .10, 1 d.f., not significant; Negro/Anglo χ^2 = .02, 1 d.f., not significant; Mexican-American/Negro χ^2 = 1.37, 1 d.f., not significant.
1961: Negro/Anglo χ^2 = 2.63, 1 d.f., not significant.

FEMALES:
1959: Mexican-American/Anglo χ^2 = 25.15, 1 d.f., p $<$.001.
1960: Mexican-American/Anglo χ^2 = 49.03, 1 d.f., p $<$.001; Negro/Anglo χ^2 = 7.66, 1 d.f., p $<$.01; Mexican-American/Negro χ^2 = 19.36, 1 d.f., p $<$.001.
1961: Negro/Anglo χ^2 = 3.49, 1 d.f., not significant.

found among either Anglos or Negroes. Mexican-Americans, at least at the time of the survey, perceived a tavern as a "man's place," frequented by "evil women." Both Anglo and Negro wives attended taverns more than Mexican-American wives, and Anglo wives more than Negro. In none of the three years of the study was there a significant difference between ethnic groups with respect to tavern attendance by husbands, but Anglo females seemed far more reluctant than Mexican-American females to report such behavior on the part of their husbands. It would seem that the Mexican-American female accepts the tavern attendance of her husband and his often rather completely separate social life centered in a tavern with friends as part of the social role of the Mexican-American male.

The proportion of male Mexican-Americans (48 percent) and Negroes (56 percent) who fished and hunted was considerably below that of the Anglos (64 percent) in 1960 (and other years as well) but Negro wives (38 percent) hunted and fished significantly more than either Anglo (26 percent)or Mexican-American (20 percent).

When respondents were asked if they participated in sports and

TABLE 57. SOCIAL PARTICIPATION SCALE, 1960 (IN PERCENTAGES)

Scale Type	Mexican-American			Negro			Anglo		
	Total	Male	Female	Total	Male	Female	Total	Male	Female
MALES:*									
0 Low participation score	25	17	34	25	23	27	15	9	22
1	4	4	3	10	8	12	5	6	5
2	8	7	10	10	9	10	7	7	6
3	12	14	11	5	4	6	7	8	6
4	10	9	11	13	15	10	9	11	7
5	32	39	26	29	30	27	29	30	28
6	7	9	5	8	9	7	15	16	15
7 High participation score	1	2	1	0	0	1	13	14	12
	99	101	101	100	98	100	100	101	101
FEMALES:**									
0 Low participation score	24	22	26	28	24	31	17	18	15
1	21	17	25	18	21	15	10	11	9
2	23	23	22	8	11	5	11	13	10
3	0	0	0	0	0	0	0	0	0
4	0	0	0	0	0	0	0	0	0
5	26	30	22	35	31	38	38	40	36
6	4	5	2	11	12	10	10	10	10
7 High participation score	2	2	2	1	1	1	14	9	20
	100	99	99	101	100	100	100	101	100

Males: R = .8661; MR = .6163; Mexican-American/Anglo χ^2 = 14.47, 1 d.f., p < .001; Negro/Anglo χ^2 = 22.39, 1 d.f., p < .001; Mexican-American/Negro χ^2 = .40, 1 d.f., not significant.

Females: R = .8048; MR = .6688; Mexican-American/Anglo χ^2 = 46.93, 1 d.f., p < .001; Negro/Anglo χ^2 = 13.11, 1 d.f., p < .001; Mexican-American/Negro χ^2 = 11.41, 1 d.f., p < .001.

*Male, as reported by male; female, as reported for male by female spouse.
**Female, as reported by female; male, as reported for female by male spouse.

[246]

hobbies, in 1959 and 1960 Anglos responded in the affirmative to a significantly greater extent than Mexican-Americans when reporting either the activities of husbands or wives. In 1960 the difference between Anglo and Negro husbands was not statistically significant but Anglo wives participated significantly more than their Negro counterparts. Negro husbands and wives participated significantly more than Mexican-American husbands and wives. In 1961, Anglo husbands and wives engaged in "participant" sports significantly more than Negro husbands and wives but there was little difference between Negro and Anglo husbands in "spectator" sports. More Negro than Anglo wives watched and significantly so.

Respondents were asked in the 1961 survey if they attended labor union meetings; about 80 percent of the Negroes in comparison to about half the Anglos applied in the affirmative. One would expect (with occupational level uncontrolled) a large proportion of industrially employed Negroes to respond in this manner in comparison with an Anglo sample that was skewed toward the middle and upper end of the occupational level continuum where union membership would not be a prerequisite to employment.

Social Participation Scales

The various 1960 items on social participation were combined in a scale for males and females. The person who scored highest on the male social participation scale went fishing or hunting, went to movies, went to taverns, participated in sports or hobbies, went to dances or parties, went to clubs, and participated in church organizations. The female who had the highest social participation score went to movies, went to dances or parties, participated in sports or hobbies, went to taverns, went fishing or hunting, participated in church organizations, and went to clubs. The distribution for both scales is presented in Table 57. Anglos had higher male and female scores than did Mexican-Americans. Negro females had higher social participation scores than either Mexican-American or Anglo females. It should be noted that the significant differences to which we refer on each scale are based on social participation rates for males as reported by themselves and males as reported by their spouses. A similar table was constructed for the females. The question arises, of course, as to whether or not spouses report accurately on each other's participation. This question cannot be answered but we can compare their reports in order to ascertain the extent to which they are congruent.

We found that males in any ethnic group described themselves generally as more active than was indicated by their spouses. The only statistically significant difference that appeared between male and female reporting was that by Mexican-American females of their spouses' participation which may be based either on underreporting of females or overreporting of males.

Conclusion

It is difficult to assess the differences in associates and association between Mexican-Americans, Negroes, and Anglos without considering social class or socioeconomic status. The data on the whole indicated that more Anglos than Negroes or Mexican-Americans, regardless of age, participated in all social activities, even those which required little or no expense. The low levels of social participation described for Mexican-Americans were in the areas expected.

The literature has quite consistently described a low rate of participation in voluntary associations as a lower socioeconomic characteristic, but such behavior is not to be confused with "social isolation." The Mexican-American youth whose activities center in the "palomilla" is not isolated but highly integrated.[18] Similarly, "compedrazgo," as it originally functioned in the Southwest or as modified today, may provide a channel or pattern for interaction in place of the more formal, voluntary, secondary group patterns that characterize the larger Anglo society.[19] Moreover, in Racine, as in other urban, industrial communities, social interaction for the urban "blue-collar" or working class person typically centers around relatives and the extended family, or athletic clubs organized around the neighborhood tavern. For example, some taverns in Racine sponsored baseball teams to which both Anglos and Mexican-Americans belonged.[20]

The crucial importance of the nature of patterns of association was hypothesized at the inception of the study. As detailed data were collected on the social characteristics of the associates of the respondents and the participation of respondents in group activities, racial or ethnic differences appeared to be so marked that the correlation of associations with various measures of economic absorption and cultural integration (as will be seen in a later chapter) must be viewed as of utmost importance. Ethnicity, race, and occupational level are powerful determinants of the people with whom one will interact and the place in the community where this interaction will take place. Knowledge of the nature of the associations of Mexican-Americans or Negroes, particularly when departing from the norm, may well become our most powerful predictor of who will be absorbed into the economy and who will be integrated into the larger culture within each of the race or ethnic groups.[21]

NOTES

1. It may appear to some readers that an error has been made in describing the response pattern of those who have the closest ties to their former home, since one of the statements, "did not plan to go back to former home to live," seems to indicate an affinity for Racine rather than former home. But many of those who failed to keep in touch with relatives and who were most satisfied with Racine *did* plan to go back to their former home to live at some future date. This seems to indicate a relationship between conception of oneself as successful and the ability to keep in contact with former home. The migrant who keeps in touch with his home is in many cases the one who goes back periodically during difficult periods in Racine and who is less satisfied with Racine but for one reason or another is determined to make Racine his home. A careful analysis of the relationship of ties to former home and other variables is necessary.

2. A second scale was constructed measuring ties to former home, which emphasized the degree of orientation toward the Southwest among those Mexican-Americans who were interviewed for the first time in 1960. The respondent most oriented toward the Southwest had the following characteristics: (1) most of husband's family lived in southern U.S. or in Mexico, (2) most of wife's family lived in southern U.S. or in Mexico, (3) the closest person outside of Racine that the family keeps in touch with lived in southern U.S. or Mexico, and (4) the second closest person that the family keeps in touch with lived in southern U.S. or Mexico. Since respondents were distributed across the scale with the following percentages in each scale type: 32, 15, 16, and 25, rather than a skewed or otherwise unbalanced distribution, it will be utilized in a later chapter where measures of orientation are correlated with measures of economic absorption.

3. But Grebler, Moore, and Guzman, *The Mexican-American People,* New York: The Free Press, 1970, p. 355, have pointed out that their own survey research data and other studies show a decline in "family communism" with increasing acculturation and social mobility. Their figures on aid given and received in Los Angeles and San Antonio suggest that more assistance has been given to relatives at an earlier period than is taking place in Racine.

4. Pertinent to our findings are those reported by Ted T. Jitodai in "Migration and Kinship Contacts," The Pacific Sociological Review, Vol. 6, No.

2, Fall, 1963, pp. 49-55. Jitodai reports that ". . . contact with relatives is the most frequent type of contact among all residents in Detroit. Contacts with neighbors and friends appear to be the next most frequent, and contacts with co-workers appear to be substantially fewer than the others." Jitodai goes on to say that "Such a pattern persists when the relationship is analyzed in more detail by considering migrant status, socio-economic status, sex, and region. . . . With few exceptions, a greater proportion of the respondents, irrespective of sex, socioeconomic status, urban-rural or regional backgrounds, have more weekly contacts in kinship groupings than in any other informal groupings. The one exception to this pattern occurs among white collar urban migrants who have a greater proportion of weekly contacts with friends than with relatives. Weekly contacts with neighbors and friends again appear to be the next most frequent, and the proportion of weekly contacts with co-workers among the respondents appears to be substantially lower than the others."

5. Although Mexican-Americans in San Antonio had predominantly Mexican-American associates in every type of formal and informal category of interaction, there were differences related to income. Higher income persons at least as frequently or more frequently had Anglo associates than lower income persons. Over half the Mexican-Americans in San Antonio for the period 1965-1966 stated that all of their present friends were Mexican-American while only 40 percent reported that all of their children's friends were Mexican-American. Within income groups even greater differences were found, depending on whether one resided in the ethnic community or outside of Mexican-American neighborhoods. Grebler, Moore, and Guzman, op. cit., pp. 396-397.

6. See Paul B. Sheatsley, "White Attitudes Toward the Negro." in *The Negro American,* Talcott Parsons and Kenneth B. Clark (eds.), Boston: Beacon Press, 1966, pp. 303-324. Sheatsley describes the pro-integration sentiments of a national sample of white people interviewed by NORC in December, 1963. The questions ranged from those which were most favorably responded to, "Do you think Negroes should have as good a chance as white people to get any kind of job, or do you think white people should have the first chance at any kind of job?" to which positive responses were 82 percent and at the other extreme, "Negroes shouldn't push themselves where they're not wanted," to which 27 percent disagreed slightly or disagreed strongly. In between was the question, "How strongly would you object if a member of your family wanted to bring a Negro friend home to dinner?" to which 49 percent said, "Not at all." Sheatsley, of course, found that mean scores on the pro-integration scale varied largely from region to region, from rural to urban, and so on. In the North the stronger the respondent's religious beliefs, the higher his pro-integration scale. On a four-point scale, however, those in the South with very strong religious beliefs had the lowest mean pro-integration scores, while those with strong religious beliefs had the highest pro-integration scores. Pro-integration scores were also positively correlated with educational level, family income, and occupational level.

7. Whether the Negroes wished to integrate with the Anglos may be irrelevant in the sense that Anglos have expressed their disapproval of engaging in similar behavior with Mexican-Americans and Negroes. For example, Alphonso Pinkney reports in "Prejudice Toward Mexican and Negro Americans:

A Comparison," *Phylon*, Winter, 1963, pp. 353-359, as reprinted in *Mexican-Americans in the United States*, John H. Burma (ed.), Cambridge: Schenkman Publishing Company, 1970, that residents of a Western city of 100,000 were asked if they would approve of situations ranging from living in integrated neighborhoods with Mexican-Americans and Negroes to hiring them as department store clerks. Approval for Mexican-Americans ranged from 45 percent for the first to 76 percent for the latter but from only 23 percent to 52 percent for Negroes. On a question as to whether Mexicans and Negroes should have the right to mix socially with other Americans, 46 percent responded "yes" for Mexican-Americans and 52 percent for Negroes.

   8. Gary T. Marx, *Protest and Prejudice: A Study of Belief in the Black Community*, New York: Harper & Row, 1969, p. 186.

   9. Ibid., pp. 190-191.

   10. Ibid., pp. 192-193.

   11. The problem of Mexican-American leadership is discussed in James B. Watson and Julian Samora, "Subordinate Leadership in a Bicultural Community: An Analysis," *American Sociological Review*, Vol. 19, No. 4, August, 1954, pp. 413-421. The authors point out the dilemma of the Mexican-American very well when stating, ". . . that the very traits which would qualify an individual to provide the sort of leadership called for are sure as to cast suspicion upon his loyalty in the eyes of many he would lead. Is it that the qualified 'leaders' make little effort to lead effectively because they feel—perhaps correctly—that they would have difficulty in getting an effective followership? Or is it that they get no effective following largely because of their own reluctance to exert leadership?"

   12. There is, of course, a large literature on the relationship of income, education, occupation, and so on, to social participation. See, for example: William G. Mather, "Income and Social Participation," *American Sociological Review*, Vol. 6, No. 3, June, 1941, pp. 380-383; Floyd Dotson, "Patterns of Voluntary Association among Urban Working-class Families," *American Sociological Review*, Vol. 16, No. 5, October, 1951, pp. 687-693; Basil G. Zimmer, "Participation of Migrants in Urban Structures," *American Sociological Review*, Vol. 20, No. 2, April, 1955, pp. 218-224; John J. Foskett, "Social Structure and Social Participation," *American Sociological Review*, Vol. 20, No. 4, August, 1955, pp. 431-438; John C. Scott, Jr., "Membership and Participation in Voluntary Associations," *American Sociological Review*, Vol. 22, No. 3, June, 1957, pp. 315-326; Charles R. Wright and Herbert H. Hyman, "Voluntary Association Memberships of American Adults: Evidence from National Sample Surveys," *American Sociological Review*, Vol. 23, No. 3, June, 1958, pp. 284-294; Robert W. Hodge and Donald J. Treiman, "Social Participation and Social Status," *American Sociological Review*, Vol. 33, No. 5, October, 1968, pp. 722-740.

   The research cited in this note leads one to conclude that social participation of various types increases with socioeconomic status and time in the community. In other words, so-called working class people and newer arrivals to the community should not be expected to indicate that they belong to clubs and other organizations or participate in the activities of clubs and other organizations to the same extent as middle class persons and those who have always resided in the community. With time in the community, migrants become more like their hosts.

13. So far as we have been able to see, the development of specialized Negro social organizations in Racine had not begun to reach the level described for Chicago by St. Clair Drake and Horace R. Cayton in *Black Metropolis,* New York: Harper & Row, 1962. See pages 669-710 for a detailed description of organizational life of the middle class Negro in Chicago.

14. Morris Axelrod, "Urban Structure and Social Participation," American Sociological Review, Vol. 21, No. 1, February, 1956, pp. 11-18.

15. In reference to the participation of Negroes in voluntary associations, see Marvin E. Olsen, "Social and Political Participation of Blacks," American Sociological Review, Vol. 35, No. 4, August, 1970, pp. 682-697. Olsen's research suggests that when socioeconomic status is controlled Negroes are more active than whites. He states, ". . . when the recent data from Indianapolis are compared with earlier data from Detroit, a clear time trend emerges. In most of the social and political areas examined in Detroit, blacks still participated less actively than whites after socioeconomic status was held constant. In contrast, in all 14 areas investigated in Indianapolis, blacks generally tended to participate more actively than whites of comparable socioeconomic levels. To the extent that we can draw inferences from this comparison of two different cities, this time trend would suggest that the civil rights efforts of the 1960's may have given blacks increased impetus to take part in all kinds of social and political activities." Olsen continues to state that ". . . the finding that blacks who identify as members of an ethnic community tend to participate more actively in most areas than do nonidentifiers (both without and with socioeconomic status controlled) also supports the ethnic community explanation. If we are correct in assuming that ethnic identifiers look to the black community as a reference group more than do nonidentifiers, and that the currently prevailing norms of the black community stress activism, then the higher rates of activity among identifiers suggest that membership in a cohesive ethnic community does propel many individuals toward participation in a variety of social and political arenas."

Also see Nicholas Babchuk and Ralph V. Thompson, "The Voluntary Associations of Negroes," American Sociological Review, Vol. 27, No. 6, October, 1962, pp. 647-655. Their study of Negro affiliations in Lincoln, Nebraska, showed Negroes to be more affiliated with formal voluntary associations than whites, especially at the lower class level. They suggest that this may be explained by Negro exclusion from other sectors of organized life in American society and that the organization of Negroes is a generation behind the general American pattern. A voluntary association provides the Negro with an opportunity for expression and recognition that he cannot achieve elsewhere in American society.

16. Although we have so few Protestant Mexican-Americans to preclude any detailed analysis of them, the reader may wish to turn to Chapter 20, "Protestants and Mexicans," of Grebler, Moore, and Guzman, op. cit., pp. 486-512.

17. For an excellent article on religion and the Negro in the United States, see Joseph H. Fichter, "American Religion and the Negro," in *The Negro American,* op. cit., pp. 401-422.

18. We have already mentioned the palomilla as a network of informal dyadic relations between age-mates as described by Arthur J. Rubel in *Across the Tracks*. Although Mexican-Americans do join voluntary associations, Rubel has pointed out that they participate in them as if they were of a primary nature. Rubel contrasts Mexican-American participation in groups with Anglo multiple membership in secondary associations of a formal character. He states that until Mexican-Americans learn to organize their social behavior in this way they cannot expect to be effective in the social, economic, and political systems of the total society. His contention is that the Mexican-Americans act as if the larger society was organized on the basis of small family units and as a consequence are unable to control much more than the social environment within the bounds of their family. Arthur J. Rubel, *Across the Tracks: Mexican-Americans in a Texas City*, Austin: University of Texas Press, 1966; specifically see Chapter 6, "Formal Organizations."

19. Grebler, Moore, and Guzman, op. cit., pp. 354-355, comment on *compadrazgo* at some length. They conclude from their own research and a careful analysis of the literature that compadrazgo is probably a minor feature of kinship and community social organization in urban centers today. They conclude, as do we, that casual use of the term *compadre* among Mexican-Americans may make the institution conspicuous, even though it no longer has the real importance that it once did.

20. Floyd Dotson, op. cit., p. 690.

21. Litwak has pointed out that the voluntary association permits strangers to be integrated into larger societies faster than most other forms of interaction, since the initiative for joining the organization can generally come from either the stranger or the native. Litwak correctly points out that in any friendship relationship the stranger would be considered "pushy" if he initiated a personal contact, and as a consequence is better able to make contacts in a voluntary association than in other types of voluntary social relations. Eugene Litwak, "Voluntary Associations and Neighborhood Cohesion," American Sociological Review, Vol. 26, No. 2, April, 1961, pp. 258-271.

Chapter 11

PERCEPTIONS OF COMMUNITY WELFARE AGENCIES

AND ORGANIZATIONS

Introduction

Directors of social service agencies understandably prefer to believe that the services of their organizations are essential to the welfare of the total community or at least to a particular segment of it. At the same time, they probably are aware that their perceptions of the role of an agency might not correspond to either that of persons in the larger community or the intended recipients of the services offered by it. An agency's clientele may be quite critical if they have not received anticipated services or treatment. Some in the community may have a favorable or unfavorable image of it, based on what they have heard from friends who have experienced a contact; others have a completely negative attitude toward all welfare agencies.

Patricia Cayo Sexton has commented on the role of the social worker. As she puts it,

"The social worker is an easy target, sometimes too easy. Like other 'public servants' [the teacher, policeman, etc.], the social worker is an alien emissary of the middle class world, put into the slum to do good and keep order. Usually he is not doing much to correct poverty—nor to create it either. The social worker did not make the slum; still he is sometimes resented for his patronage and abused because he is close at hand."

She goes further in quoting Marion K. Sanders in Harper's Magazine,

"The day after the bomb fell, the doctor was out binding up radiation burns. The minister prayed and set up a soup kitchen in the ruined chapel. The policeman herded stray children to the rubble heap where the teacher had improvised a classroom. And the social worker wrote a report; since two had survived, they held a conference on Interpersonal Relationships in a Time of Intensified Anxiety States."[1]

Correctly or incorrectly, the social worker is the target of the less fortunate who, at least in part, look about for individuals whom they can blame for their misery. Some militant elements in the ghetto now perceive the organization of the larger society as being a more appropriate target, although they continue at times to turn their aggressions upon individuals whom they perceive to be representatives of the larger middle and upper class society.

In an earlier article, various alternate explanations of the images of social and welfare agencies were presented at some length. Perhaps we were more sympathetic at the time than is currently the vogue.

"Unfortunately, welfare agency personnel often either develop resistance to finding out what kind of an image their agency has in the community or, should someone detect that the image is an unfavorable one, they fail to seriously consider how this image happened to develop. Probably the first reaction of agency personnel, should it be found that the image is unfavorable, is one of defense and rationalization. It is easier to take the view that a valuable function is performed, but that it is unappreciated by the community, than it is to conceive of the agency as not performing a valuable function for the community. Hence it is sometimes decided that an educational campaign is necessary in order that the community better understand the purpose and function of the agency. On the other hand, if an agency is not of much assistance to the less fortunate persons in the community, that may be less a function of the manner in which the agency is operated than of the fact that the community's resources are inadequate, or that there really is no easy solution to some of the problems with which the community is confronted."[2]

In 1961, with what we considered to be noteworthy concern, various Racine social service agencies requested that we ascertain perceptions of respondents of the following community organizations: The Red Cross, the Salvation Army, the Society of St. Vincent de Paul, Racine County Welfare, the National Association for the Advancement of Colored People, the Mayor's Commission on Human Rights, the Community Service Council, the Urban League, and the Muslim Temple. A fictitious agency was also included in the schedule as a reliability check.

The following questions were asked in order to measure different levels of contact and knowledge:

(1) Have you heard of the activities of any of the following organizations?
(2) What are your feelings regarding this kind of activity?
(3) Have you had any contact with this organization?
(4) (IF YES—HAD CONTACT) How would you describe the people with whom you have had contact?
(5) (IF NO—NO CONTACT) What have you heard about the kinds of people who are in this organization?

Besides being interested in the image of each agency in the community as projected by responses, we were concerned with whatever difference might occur between Negro and Anglo perceptions. If there were interagency and intergroup differences, would they be explained by the nature of the agencies, the nature of their target population (clients), or by the differential effectiveness of agencies in teaching and servicing various groups in the community?

The first problem involved an examination of the responses in terms of some sort of rank ordering of agencies and organizations. Five sets of ranks corresponding to the five questions presented above could be made. Although the agencies and organizations were ranked according to either Negro or Anglo responses, we recognized that the perceptions of the Anglos constituted in reality those of the larger community, since in 1961 Racine was more than 90 percent Anglo. Because our concern was centered on the people serviced by agencies, the viewpoint of Negroes (and Mexican-Americans) assumed increased importance in the process of evaluation, albeit only Anglos and Negroes were actually interviewed. Rather than present an over-all evaluation of agencies, we have emphasized Negro versus white perceptions in the analyses that follow.

Recognition of Agencies and Organizations

The nationwide activities of an organization, the degree of national attention focused on its activities, the emphasis of the organization on public relations, the length of time that an organization has been providing services, and the type of service it renders (direct client versus community service) are factors which can qualify or influence the degree of recognition which will be accorded an agency or organization. The response data to the five questions, based on the rank order of Negro responses to the first question, are presented in Table 58. These data indicate that organizations such as the Red Cross and the Salvation Army were quite well known to either Negro or Anglo respondents. while others were scarcely acknowledged. The most noticeable difference between Negroes and Anglos was in disparate recognition of the NAACP and the Muslims by the Negroes and the St. Vincent de Paul Society and the Community Services Council by the Anglos

TABLE 58. CHARACTERISTICS OF SOCIAL AGENCIES AND COMMUNITY ORGANIZATIONS ACCORDING TO RACE OF RESPONDENTS, PERCENT RESPONDING POSITIVELY, 1961

	Heard of		Favorable Feelings		Had Contact		Favorable Perception Contact:		No Contact:	
	%	Rank	%	Rank	%	Rank	%	Rank	%	Rank
NAACP										
Negro	96	1	90	1	45	2	77	3	68	1
Anglo	76	5	44	6	3	7.5	17	7	34	4
Salvation Army										
Negro	93	2.5	76	3	34	3	85	1	36	3
Anglo	97	1.5	81	1	21	3	90	1	55	1
Racine County Welfare										
Negro	93	2.5	65	4	47	1	40	5	34	4
Anglo	85	4	73	3	23	2	53	6	37	3
Red Cross										
Negro	91	4	78	2	26	5	75	4	37	2
Anglo	97	1.5	56	5	26	1	62	5	33	5
Muslim Temple										
Negro	70	5	11	9	28	4	32	6	11	8
Anglo	12	9	9	9	1	8.5	—*		0	
St. Vincent de Paul										
Negro	50	6	62	6	17	6	78	2	33	5
Anglo	86	3	80	2	11	4	81	2	44	2
Urban League										
Negro	28	7	50	7	1	8	—		6	9
Anglo	21	8	26	8	1	8.5	—		3	8
Mayor's Commission on Human Rights										
Negro	27	8	64	5	1	8	—		30	2.5
Anglo	40	6	43	7	3	7.5	67	4	25	6
Community Service Council										
Negro	11	9	46	8	1	8	—		30	6.5
Anglo	34	7	69	4	7	5	78	3	24	7

*A dash (—) indicates insufficient contact to present a percentage in this column.

Attitudes Toward the Activities of Agencies and Organizations

The Negro sample reacted most favorably toward the NAACP and the Red Cross, whereas the Salvation Army and St. Vincent de Paul received the highest percentage of Anglo favorable responses. The Muslim Temple received the least approval from either group.[3] Negroes had a more favorable perception than Anglos of the Mayor's Commission on Human Rights and the Urban League, while the Anglos favored the Racine County Welfare and the Community Service Council more than did Negroes.

We did not expect all social welfare agencies to be recognized by our respondents, nor did we expect a particularly favorable evaluation of those that were known. Considering the literature, the responses of our samples could quite conceivably have been more critical than they were. Arthur B. Shostak has commented on this problem in some detail:

> "The serious deficiencies of welfare allowances are compounded several times by the failures of various welfare agencies. Some of the agencies fail to integrate their work with that of related agencies; many have inadequate personnel; caseworkers are often overworked; the voice of the client is not always heard in policy-making agency councils; and the philosophy of the entire service effort is woefully misguided and self-defeating.

> "To begin with, 'hard-to-reach' clients are frequently the product of 'hard-to-reach' agencies. 'Multiproblem families' may be a reflection of 'single-purpose agencies'; and in slum neighborhoods, these agencies frequently operate so blindly that no continuing and reinforcing contacts exist among them."[4]

Technical restrictions placed upon welfare clients vary from state to state and community to community and, in some places, actually handicap the client who is seeking work. Additional restrictions are of such nature that the client perceives them as indignities beyond those which he already feels because he must apply for welfare. On this point Shostak elaborates:

> "Second, agency personnel are sometimes inadequate because their commitment is to a narrow psychoanalytic framework, one that leads them to condemn and censor the client for his presumed personal faults. The poor, Irving Kristol [Irving Kristol, "Poverty and Pecksniff," The New Leader, March 30, 1964, p. 23] reminds us are 'entitled to have every possible opportunity to move upward on our socioeconomic scale, if they wish to. But they are also entitled not to be hectored, badgered, sermonized, psychoana-

lyzed, fingerprinted, Rorschached, and generally bossed around by a self-appointed body of self-anointed redeemers. . . . '"[5]

On the other hand, it is evident that the community usually does not provide public welfare agencies with sufficient funds to carry on a program at the level that its directors deem appropriate. Few agencies ever have a sufficient number of case workers to adequately care for their clients. Shostak mitigates his criticism to some extent by pointing out that:

"... caseworkers, whatever their theoretical framework, seldom have adequate time and energy to spend with all their clients. A social worker in New York State, for example, must fill out 24 separate pieces of paper before a client gets his first welfare check. Sixty-four forms are labeled 'most frequently used.' Social workers are given case loads well in excess of the recommended assignment of seventy-five clients (150 to 200 are common). Little individual attention is possible; little real aid is given. Overwork, combined with extremely low salaries, helps to produce a very high rate of turnover (nearly 50 percent of the staff turns over annually in several large industrial states). As usual, there is little continuity in the help extended to a given family."[6]

Contact with Agencies and Organizations

The orientation of social work in the United States in recent years lends little credence to the belief that most agencies have contact with the poor. The troubled middle classes whose problems are perhaps more amenable to individual treatment become much better targets for social work agencies inasmuch as work with them is not only likely to be more pleasant and personally satisfying but the nature of the troubled middle class client's problems make it easier for the agency to report that progress has been made than when a similar amount of time is spent on the problems of the "deserving" or undeserving poor. Nathan E. Cohen, a professor of social welfare at the University of California in Los Angeles, describes the situation in the following quotation:

"This brings me to my final concern, namely, our theory and methodology for working with people in trouble. Private social work, like other helping professions, has tended to disengage itself from the poor. Theoretically, social work began with a sociological period and then, with the impact of the psychoanalytic, shifted to a heavy emphasis on individual treatment. In its sociological period, there was a strong current of social reform reflecting the influence of pragmatism with its view that man was master of his own fate, and collectively could build a brave new world. The Freudian emphasis shifted the focus toward a Hobbesian view of man, but tended to abstract the individual from the power struggle of the various social institutions of which he was a part. This created a dualism in social work thinking.

"Freud, from whom social work borrowed much, was pessimistic about the value of applying the therapeutic approach to the poor. As stated by Friedenberg (Edgar Z. Friedenberg, "New-Freudianism and Erich Fromm," Commentary, Vol. XXXIV, October, 1962, p. 309] .

"His tangibly compassionate reservation about using psychoanalysis to treat working class patients suggests that he felt that, for most individuals, the candle really did cost more than the game would be worth; and that people who in reality had very little opportunity to lead richer lives might better be left with their defenses and illusions.

"For Freud, the emphasis was on the individual and his adjustment to society. Not enough consideration was given to a changing society as well as a changing individual. Fromm has pointed this up sharply. For Fromm a pervasive socioeconomic structure which runs counter to the basic nature of man, with 'its inherent need for love, human solidarity and the development of reason' can result in a 'socially patterned defect' in the total society. Such a condition, states Fromm, 'inhibits his faculty for critical thought, and tends to transform him into an automaton, into a marketing personality who loses the capacity for genuine and profound feeling and thought, and whose sense of identity depends on conformity' [Erich Fromm, "Psychoanalysis," *What is Science?* John R. Newman (ed.), New York, Simon and Schuster, 1955, p. 380].

"The need to break out of the narrow treatment context has become evident even in the work with middle-class. It has been made most visible, however, in dealing with the problems of poverty. There is growing recognition that poverty cannot be viewed only in psychological terms, that it also has its roots in 'structural' problems in our society. Furthermore, it is not just a problem for the individual poor, it has become a challenge to the survival of our democratic society as an institution of the people, by the people, and for the people." [7]

In essence, what Cohen is saying is that individual therapy is entirely inadequate for the problems of the poor which are related to the organization of society. Not only do we agree, but we stress and emphasize a further step, namely, the extent to which the organization of the larger society makes it difficult and at least in some cases almost impossible for even the most "ambitious" inmigrant, or particularly one with a language handicap, to become absorbed into the economy at anything other than the lowest levels.

S. M. Miller has written in the same vein when he speaks of the necessity of clarifying the goals of professionals in the social services:

"Services aimed at individual treatment are not enough. Professionals and their organizations must support and encourage action that will deal with the larger American scene in which poverty is being produced and maintained. The professional role cannot end with the limited services that the profession provides; it must extend itself to pressure for social changes that will make individualized professional services more meaningful and effective. Concentrating on individualized services, without concern for the forces outside the profession that are molding and limiting possibilities, is tantamount to adopting professional blinders. These blinders may promote confidence in one's expertise and effectiveness, but they force the professional to ignore the barriers to deep and continuing change among the clientele.[8]

The proportion of each sample that had contact with various agencies or organizations is presented next in Table 58; contact did not include contributing to agencies or soliciting for funds. It was not surprising to find that almost one-half the predominantly lower socioeconomic status Negro sample had direct contact with the Racine County Welfare Department. The Anglo sample, mostly of higher socioeconomic status, had its greatest contact with the Red Cross. Those agencies which showed a significant difference in Anglo/Negro contacts were the Racine County Welfare, NAACP, Salvation Army, and the Muslim Temple. In each of the above the number of Negro contacts was much higher than the Anglo.

Perception of Agency Personnel

Whether Anglos and Negroes differ in their expectation of the services to be performed by agencies and organizations, or whether the treatment of clients varies by race, were not the issues in this study. Perception of agency personnel was based on the favorableness of responses of those in each sample who had contact and those who had no contact with them. The Salvation Army, the Red Cross, and the Society of St. Vincent de Paul received significantly more favorable responses from both Negroes and Anglos with contact as contrasted with those who had none. Significantly more Anglos with contact rather than without were favorably disposed toward the Community Service Council and the Mayor's Commission on Human Rights. The numbers of Negroes who had contact with these organizations was too small for comparison. Those Negroes who had contact with the Muslim Temple were significantly more favorable toward that organization than those who had no contact.

When those who had no contact with the agency or organization were asked what they had heard about its personnel, the favorable Anglo image of the St. Vincent de Paul Society and of the Salvation Army was greater than that of the Negro. The NAACP and the Muslims had more favorable images among the Negroes than the Anglos.

In most cases with sufficient responses for comparison, there was a significant difference between the percentage with favorable impressions who had contact and those who had no contact with agencies or organizations. Every difference, whether significant or not, was in the direction of disproportionately more respondents having a favorable perception if contact had been made with the agency. In this respect, the Anglo and Negro patterns were essentially the same.

Handler and Hollingsworth[9] found that it was not the attitudes and practices of the agency or of individual case workers but the client's feelings of embarrassment in the presence of non-welfare people which gave recipients feelings of stigma. In fact, the authors found that clients who felt stigma were less happy with their case workers in the program, tending to use the program more and ask their case workers for more. Furthermore, they left the program sooner and to a greater extent by their own efforts. The authors do not infer that stigma is a desirable thing, but point out that unfortunately people act in socially desirable ways because of feelings of shame. The more worrisome cases, according to the authors, are the recipients who do not have feelings of stigma. These are the recipients who are passive, accepting, satisfied, and unable to realize the advantages of the program. High levels of reported satisfaction and lack of complaints are not necessarily signs that things are going well nor that a program is operating effectively.

In another study, Handler[10] found that top level administrators and administrators with long experience gave an evaluation of AFDC appeals which differed from that offered by younger administrators and administrators closer to the field who stated that there was a great deal of withholding of information from clients, that administration was highly discretionary, and that the administrative appeals system was not working because clients were either unaware of their rights or were afraid of exercising them.[11]

Some of the above comments are more pertinent to the welfare and social services situation in 1960, for since that time the War on

Poverty and the growth of Community Action Programs has generated a limited amount of citizen participation in welfare policy.[12]

To continue our interpretation of Table 1, it has been suggested that not only favorable opinions but having an opinion varies by race and by contact. In other words, variation in favorableness of opinion about any agency or organization is subject to whether or not one has an opinion. When the percentages of persons with no contact and no opinions about the personnel of agencies and organizations were compared by race, there were significant differences for only three agencies or organizations—the NAACP, the Muslim Temple, and the Salvation Army. The Negroes were more likely than Anglos to have had no contact with an agency and to indicate no opinion of an agency's personnel, even though they were more often aware of the organization's existence than were the Anglos. This response pattern was such as to lend additional evidence to the hypothesis that Negroes had different sources of information for their evaluation of agencies and organizations than Anglos.

One immediately wonders how much of the difference between Negro and Anglo images is based on the fact that many Negroes have neither been absorbed into the economy nor integrated into the larger community to a sufficient extent to have concrete knowledge of the operation of its institutions. Differences between Anglos and Negroes tend to disappear among those who have had actual contact with the social welfare agencies and community organizations. It is very possible that members of the larger Anglo society obtain their information from a variety of secondary group sources while Negroes, being less integrated into the formal communication system, are most likely to derive their knowledge and opinions from communication with friends who have had contacts.

Summary Measures

Thus far we have either presented each dimension of an agency or organization separately or one dimension in comparison with another. To satisfactorily represent the image of an agency in the community with a single composite score is a somewhat difficult task because the five questions in the schedule do not constitute specific dimensions of one universe of content but at least two: (1) recognition and contact, and (2) favorableness. Despite the absence

TABLE 59. SUMMARY AND RATIO SCORES OF ALL POSITIVE NEGRO AND ANGLO
RESPONSES TO QUESTIONS ABOUT AGENCIES

A: Sum of Positive Responses

	Negro		Anglo		B: Ratio of Anglo to
	Sum	Rank	Sum	Rank	Negro Positive Responses
NAACP	376	1	174	7	.463
Salvation Army	324	2	344	1	1.062
Red Cross	307	3	274	3	.892
Racine County Welfare	279	4	271	4	.971
St. Vincent de Paul	240	5	302	2	1.258
Muslim Temple	152	6	22	9	.183*
Mayor's Commission on Human Rights	122	7	178	6	.910*
Community Service Council	88	8	212	5	1.523*
Urban League	85	9	51	8	.600*

*Ratio based on responses to questions (1), (2), (3), and (5).

of unidimensionality, we have attempted in Table 59 to combine these disparate dimensions into an overall measure of positive response. This simple additive scale is by no means a perfect representation of the community's perception of its agencies and organizations, including personnel. The percentages of positive responses for each of the questions have been totaled by race in the "A" columns. The ratio of Anglo to Negro responses is shown in column "B." The higher the ratio (i.e., above 1.000) the greater became Anglo recognition and approval of an agency or organization; the lower the ratio (i.e., below 1.000) the more the Negroes recognized and approved it.

Table 60 shows the pattern of significant differences (and the level of significance) between responses to each of the five questions. An "A" or an "N" indicates whether the percentage of positive responses was greater for the Anglo sample or for the Negro sample. All agencies were ranked according to the number of significantly different Anglo versus Negro favorable responses to questions about the agencies, commencing with agencies favored by Anglos. The Society of St. Vincent de Paul received the greatest proportion of favorable Anglo responses and the NAACP the greatest favorable Negro responses.

Tables 59 and 60 illustrate the significant racial differences in the pattern of contacts with and perceptions of various social welfare agencies and organizations in an industrial community.

Our attention now turns toward individual respondents. How did

TABLE 60. SUMMARY OF NEGRO AND ANGLO DIFFERENCES ON FIVE QUESTIONS AND RANK OF AGENCIES BY RACE

	Heard of	Favorable Feelings	Had Contact	Favorable Perception	
				Contact:	No Contact:
St. Vincent de Paul	A .001	A .01	N n.s.	A n.s.	A .001
Community Service Council	A .001	A n.s.	A n.s.	—*	N n.s.
Salvation Army	A n.s.	A n.s.	N .05	A n.s.	A .05
Racine County Welfare	N n.s.	A .05	N .001	A n.s.	A n.s.
Mayor's Commission on Human Rights	A n.s.	N .01	A n.s.	—*	N n.s.
Red Cross	A n.s.	N .001	Equal	N n.s.	N n.s.
Urban League	N n.s.	N .05	Equal	—*	N n.s.
Muslim Temple	N .001	N .01	N .001	—*	N .01
NAACP	N .001	N .001	N .001	N .01	N .001

*Dashes indicate insufficient contact to present a difference in this column.

Negroes and Anglos vary in amount of contact, knowledge, and images of social welfare agencies and organizations? Is it possible to say that there was a continuum of contact, awareness, or favorableness in perceptions of social welfare agencies and organizations in the community? Is it possible to rank people in terms of the extensiveness of their contact, awareness or favorableness of perception?

The familiarity of respondents with agencies and organizations was summarized in several Guttman scales. Separate scales for Negroes and Anglos as well as one for the combined sample are presented in Table 61. Agencies recognized by Anglos from most frequently to least frequently were found in the following order: The Red Cross, Salvation Army, St. Vincent de Paul, Racine County Welfare, NAACP, Mayor's Commission on Human Rights, Community Service Council, Urban League, Muslim Temple, and Community Action League. Negro respondents were most aware of the panorama of agencies to least aware in this sequence: the NAACP, Salvation Army, Racine County Welfare, Red Cross, Muslim Temple, St. Vincent de Paul, Urban League, Mayor's Commission on Human Rights, Community Service Council, and Community Action League. When the two samples were combined in the scaling process, agencies were listed in the following order of recognition: Salvation Army, the Red Cross, Racine County Welfare, NAACP, St. Vincent de Paul, Muslim Temple, Mayor's Commission on Human Rights, Community Service Council, Urban League, and Community Action League.

TABLE 61. DISTRIBUTION OF AGENCY RECOGNITION SCALE SCORES BY RACE, 1961 (IN PERCENTAGES)

		Negro Scale	Anglo Scale	Combined Scale Scores	
Scale Scores				Negro	Anglo
Least Knowledgeable	0	1	5	2	2
	1	1	—	1	1
	2	1	16	1	6
	3	2	34	1	6
	4	17	19	19	5
	5	27	10	10	53
	6	29	10	42	3
	7	8	3	16	5
	8	9	1	1	10
	9	4	1	4	8
Most Knowledgeable	10	1	2	1	1
		100	101	98	100

Separate Scales: [Anglo R = .9392; MR = .8310] [Negro R = .9372; MR = .8320] Negro/Anglo χ^2 = 84.57, 1 d.f., p $<$.001.
Combined Scales: [R = .9110; MR = .8120] Negro/Anglo χ^2 = 43.87, 1 d.f., p $<$.001.

The Anglo and Negro scores differed significantly on either the separate or combined scales.

A second set of scales was constructed from responses to the question, "What are your feelings regarding this kind of activity?" by those who were aware of the activities of the social welfare agencies and community organizations. Table 62 shows ten different types of persons at various points on the scale. A score of zero on the scale represents the respondent who had no feelings, negative feelings, or who had not heard of the Salvation Army, the Racine County Welfare, the Red Cross, the NAACP, The Society of St. Vincent de Paul, the Mayor's Commission on Human Rights, the Community Service Council, the Urban League, and the Muslim Temple. A score of one on the scale means that the respondent had positive feelings toward the Salvation Army, but not toward the others; the scale ascends in like order, until a score of nine is reached, at which point the respondent had positive feelings toward all agencies and organizations.

This scale enabled us to efficiently represent the differences that existed between Negroes and Anglos. The distribution of Anglo scale types was significantly different from the distribution of Negro scale types; 45 percent of the Anglos appeared in scale types zero through three, while less than 20 percent of the Negroes fell in these categories. The modal scale type for the Negroes was type four,

TABLE 62. DISTRIBUTION OF SCALE SCORES BY RACE IN RESPECT TO ATTITUDES TOWARD ACTIVITIES OF AGENCIES AND ORGANIZATIONS, 1961 (IN PERCENTAGES)

		Negro	Anglo
Negative or unfavorable feelings	0	6	6
	1	5	11
	2	3	18
	3	4	10
	4	45	5
	5	23	33
	6	12	10
	7	1	4
	8	1	3
Positive or favorable feelings	9	–	1
		100	101

[R = .8860; MR = .7520] Negro/Anglo χ^2 = 24.19, 1 d.f., p < .001.

which included the Salvation Army, the Racine County Welfare, the Red Cross, and the NAACP. The introduction of the NAACP at this point brought an additional five percent of the Anglos and an additional 4 percent of the Negroes into the scale. The modal scale type for the Anglos was type five, which included all the agencies in type four plus the St. Vincent de Paul society. The Anglo sample, as represented by scale scores, had a relatively less favorable attitude toward the activities of various agencies and organizations than did the Negro sample.

It must be concluded that social service or welfare agencies and community organizations were quite differently known and contacted in the larger Anglo community as compared with the Negro subcommunity. It was also very evident that there were great differences in the extent to which people were favorably oriented towards the activities and personnel of agencies and organizations, and that Anglos were distributed on the continuum of favorable orientation quite differently from the Negroes.

NOTES

1. Patricia Cayo Sexton, *Spanish Harlem: Anatomy of Poverty*, New York: Harper & Row, 1965, pp. 99-100.

2. Lyle W. Shannon, "The Public's Perception of Social Welfare Agencies and Organizations in an Industrial Community," Journal of Negro Education, Summer, 1963, p. 276.

3. Considering the fact that the headquarters of the Black Muslims has been in Chicago since the middle thirties and that they have probably flourished in the Chicago South Side more than in any other area, it is somewhat surprising that they were not more popular in Racine. Those who are unfamiliar with the origin, growth, and development of the Black Muslims should read C. Eric Lincoln's *The Black Muslims in America,* Boston: Beacon Press, 1961, and *The Autobiography of Malcolm X,* New York: Grove Press, Inc., 1964.

4. Arthur B. Shostak, "The Poverty of Welfare in America," *New Perspectives on Poverty,* Arthur B. Shostak and William Gomberg (eds.), Englewood Cliffs: Prentice-Hall, Inc., 1965, pp. 96-97.

5. Ibid., p. 97.

6. Ibid., p. 97.

7. Nathan E. Cohen, "A National Program for the Improvement of Welfare Services and the Reduction of Welfare Dependency," *Poverty in America,* Margaret S. Gordon (ed.), San Francisco: Chandler Publishing Company, 1965, pp. 276-298, and specifically pp. 292-293.

8. S. M. Miller, "The Disengagement of Social Workers from the Poor," *Poverty in Affluence: The Social, Political, and Economic Dimensions of Poverty in the United States,* Robert E. Will and Harold G. Vatter (eds.), New York: Harcourt, Brace & World, 1965, pp. 241-243.

9. Joel F. Handler and Ellen Jane Hollingsworth, "Stigma, Privacy, and Other Attitudes of Welfare Recipients," Stanford Law Review, Vol. 22, No. 1, November, 1969.

10. Joel F. Handler and Ellen Jane Hollingsworth, "Justice for the Welfare Recipient: Fair Hearings in AFDC—The Wisconsin Experience," The Social Service Review, Vol. 43, No. 1, March, 1969.

11. Also see by Joel F. Handler and Ellen J. Hollingsworth, "The Administration of Welfare Budgets: The Views of AFDC Recipients," The Journal of Human Resources, Vol. V, No. 2, 1970; "The Administration of Social Services and the Structure of Dependency: The Views of AFDC Recipients," The Social Service Review, Vol. 43, No. 4, December, 1969; and Joel F. Handler and Margaret K. Rosenheim, "Privacy in Welfare: Public Assistance and Juvenile Justice," Law and Contemporary Problems, Vol. 31, No. 2, Spring, 1966.

12. The success of citizen participation in urban renewal and Community Action Programs is discussed in an article by Jon Van Til and Sally Bould Van Til, "Citizen Participation in Social Policy: The End of the Cycle?" Social Problems, Vol. 17, No. 3, Winter, 1970, pp. 313-323.

For an excellent study of the role of Community Action Programs in various cities throughout the United States and their relationship to established social welfare programs and social service agencies, see *A Relevant War Against Poverty: A Study of Community Action Programs and Observable Social Change,* New York: Metropolitan Applied Research Center, Inc., 1968.

Chapter 12

CAUSAL CHAINS AND THE EXPLANATION OF ECONOMIC ABSORPTION AND CULTURAL INTEGRATION

Introduction

There has been a rather long controversy in sociology about the relative merits of hypothesis testing versus the "scattergun" approach in the collection and analysis of data. We have utilized the scattergun approach in several earlier publications where we discussed the interrelationships of a large number of the variables as measured by Tau.[1]

The danger in the scattergun approach is the tendency of the researcher to select sizeable and statistically significant correlations to explain differences between samples or segments thereof and to ignore those that are low or inconsistent with his theoretical formulation of the problem. Since it is entirely possible for statistically significant correlations (the proportion depending upon the level of significance established) to occur by chance alone, it is reasonable to expect that those selected may be chance correlations. The advantage of this approach is that the researcher is allowed an almost immediate review of his first findings, which may lead to a reconceptualization of the study or perhaps the formulation of more subtle questions that will elicit new and more comprehensive data. On the other hand, examination of a matrix of correlations may suggest recoding some of the data to better account for a greater proportion of the variance in the dependent variables.

What the Scattergun Approach Revealed
in the Racine Study

As the Tau coefficients of correlation were computed for the data in 1959, 1960, and 1961, we found that the interrelationship of variables declined or changed direction when ethnicity or race was controlled. Mobility was clearly influenced by differences in the organization of racial, ethnic, or social class subcultures, the latter defined by occupation, income, and residence, or by the attitude and behavior of persons in the larger society toward the members of the specific subculture.[2]

In an effort to summarize work for the three years, all of the variables and scales were subsumed under 16 different groups. The 31 "independent" or "intervening" variables were placed in the following categories: measures of urbanization, orientation of a regional nature, occupational antecedents, occupational level of associates, educational background, social participation, handicap scales, and demographic data. Thirty-eight dependent variables or "indicators" of economic absorption and cultural integration were categorized as: measures of satisfaction with adjustment, measures of integration into the new community, solidarity of ties to former home or subculture, convergence of attitudes with those predominant in the larger community, educational and occupational aspirations, income and occupational status, level of living, and world view. The number of statistically significant correlations as compared to the total number of correlations within each cell in a matrix based on these groups were fewer than expected in most cells but more than expected in others.[3] The results showed that the Anglos occupied positions in the social order consistent with their initial advantages. The upward mobility of Mexican-Americans and Negroes was in reality an artifact of their movement from rural to urban or South and Southwest to the northern industrial community. Furthermore, occupation, income, level of living, and level of aspiration were more frequently intercorrelated for the Anglos than for either the Mexican-Americans or Negroes.

When the samples were divided according to time in the community, correlations among those in the community 10 years or more were in the direction expected and higher than for persons who had been in the community less than 10 years.[4] Nevertheless, while Negroes from the South or Mexican-Americans from the Southwest who had moved to Racine became increasingly absorbed into the industrial economy over a period of time, most telling is the fact that none of the correlations between independent and intervening variables (both antecedents of measures of absorption and integration), was higher than correlations between ethnicity or race and measures of absorption and integration. It was possible to discern that Racine was a community structured along race or ethnic lines and occupational levels. Though the structure of the community was not completely rigid, its organization was the crucial variable in determining what one had and with whom one would associate. The data clearly demonstrated that there was more variation in the community based on differences between groups than within groups.

Predicting Economic Absorption and Cultural Integration

It has often been said that science has three functions: to enable us to understand, to predict, and to control. We have attempted to describe some aspects of how the society operates. Our next question was: could we use data (antecedent, independent, or intervening variables) to predict the position of an Anglo, Negro, or Mexican-American on any measure of economic absorption or cultural integration? A sizeable and highly significant relationship was found between many variables and measures of absorption and integration but, more often than not, the distribution of the marginals precluded prediction or resulted in predictions no better than could be made from the modal category of the marginals.

For example, when the combined sample of Anglos, Mexican-Americans, and Negroes was examined [5] without holding ethnicity or race constant, we found relatively high correlations between the measures of economic absorption and cultural integration and a set of scale scores summarizing background data.[6] However, when the combined sample was divided on a basis of race and ethnicity, the significance of the association declined and scale scores did not enable us to predict occupational level with less error than could be predicted from the modal category of the marginals. Similarly, when we attempted to predict an active or passive world view of the combined samples from a scale score based on antecedent experiences, present status, and present associations, there was a 39 percent increase in predictive efficiency over the modal category of the marginals, but when the respondents were partialled by race and ethnicity the correlations declined and no increase in predictive efficiency took place.

Further attempts to predict were made controlling for time in the community. Respondents were divided into three categories: (1) those who had always lived in Racine, (2) those who had been there less than 10 years; and (3) those who had not always lived in Racine but had been there 10 years or more. Antecedent experiences were of limited use in predicting occupational status of associates. In the case of Anglos who had always lived in Racine, scale scores representing antecedent experiences allowed a 17 percent increase in predictive efficiency over the modal category of the marginals in predicting occupational level of present associates. In the case of Mexican-Americans who had resided in Racine less than 10 years, antecedent experiences and occupational level of present associates presented a 33 percent increase in predictive efficiency over the modal category of the marginals.[7]

On the other hand, when the world view of respondents, partitioned according to time in the community, was correlated with scale scores based on antecedent experiences, present occupational status, and present occupational status of associates, the correlations ranged from .2050 to .508. Prediction errors were reduced by .4705 percent for Mexican-Americans who had lived in Racine more than 10 years ago.

Some Life Histories of Inmigrants

At this point we believe it is best to present a very few brief life histories of representative inmigrant families and a general indication of their place on various scales. These sketches will indicate the complexity of the problem of predicting economic absorption and cultural integration from antecedent and experiential characteristics of the inmigrants. The first three are Mexican-American and the others, Negro; all names are fictitious.

An Older Mexican-American Inmigrant Less than
10 Years in Racine

Born in 1912, Juanita and her husband called Waco, Texas (1960) population 97,808) their hometown. They had moved from Texas, via Mississippi, to Somers (near Racine, Wisconsin), where they had worked on farms for six years. After coming to Racine in 1951, Juanita's husband obtained employment as a chipper (industrial laborer) in an iron foundry. He held the same job nine years later, at the time of the interview. His hourly wage in 1960 was $2.58. If he worked 32 hours per week, 52 weeks of the year (as he had in the previous year), his wages for the year would total approximately $4,300.

Though their family income was low and prices were high, Juanita believed that they were better off in Racine because the summers were not so hot, jobs were more plentiful, and the pay better than in Texas.

Juanita's antecedent handicap, based on her family background, education, and experience, was at the highest point on the scale. Her father had been an agricultural laborer; she had not known her grandparents but they had lived in Mexico. During the "depression years" in Texas, she had worked in a nut factory and in a government program sewing clothes for the poor. In Wisconsin, she and her children, when not in school, sometimes went to the farms to work.

Juanita felt schooling was very valuable for their nine children as a preparation for jobs which paid better salaries; without it in the modern world (1960), it would be impossible to get a job. Yet Juanita said she would be satisfied if the children received a ninth grade education, thus placing herself at the lowest point on both the aspiration for children and level of satisfaction with education for children scales.

Juanita and her husband spoke Spanish to each other and with their friends, while the children used English as well as Spanish. Her answers to the world view questions gave her almost the lowest passive score. While the interior and exterior of their house were disorderly and in need of repair, the fact that they owned it, had an old model car, and had other household items gave them a high possessions score. Juanita seldom participated in social activities and her husband's score was not much higher.

Juanita and her family are very typical of the older Mexican-American inmigrants who moved to Racine around the late forties or early fifties. Most of them have been absorbed into the economy at the lower levels.

A Younger and Very Recent Mexican-American Inmigrant

Antonio is a younger man who had been in Racine for about two years; he represents an inmigrant whose absorption into the economy was precarious. We describe his story as follows:

Antonio's hometown was San Antonio, Texas (1960 population 587,718) and also where he received his high school education. Born in 1921, he had held various jobs such as farm work as a boy, sorting stock, garage attendant, and pouring and molding steel. He also owned and managed a restaurant for several years. While on a visit to his relatives in Racine in 1958, he decided to remain there with his new wife who had been a waitress and an agricultural laborer. He found that he liked the smaller city of Racine; the people were friendlier and the job opportunities were better.

His level of aspiration for his two children was midway on the scale, neither at the highest nor at the lowest level. The children were free to select a career, though he did hope it would be professional, such as doctor. He wanted a college education for either boy or girl, but would be satisfied if he could afford only two years. His world view score was in the passive range, with quite well-formulated beliefs. Though his job in 1959 as a sandblaster had brought in over $4,000, the income from his present occupation as a janitor was around $2,900. His possessions score, on the other hand, was just above the average. He and his wife did go to dances, parties, and movies but not to taverns. They belonged to a private social club, and he liked to fish. His score was the highest possible, hers just below average on the social participation scale. The house interior was orderly though the furniture had that worn look from continual hard use. The exterior needed paint and the grass was uncut.

Antonio's antecedent handicap score was just above the mid-point of the scale. He was not only cooperative in the interview but he had been waiting for the interviewer and hoped that "it would be of value to science and knowledge." His first interview was in good Spanish; his second in English. His language score indicated high utilization of English, both English and Spanish being spoken in the family and with friends.

Antonio's case is particularly interesting because his high school education, past industrial experience, and English language facility had not enabled him to become absorbed into the economy at a higher level than persons without these advantages.

Another Older Mexican-American Inmigrant
10 Years in Racine

The interview with Estralda was conducted in Spanish. She and her husband were born in Pearsall, Texas, but considered Cotulla

(1960 population 3,960) their hometown. They came to work in the fields around Racine in 1948 and because of what people told them of job opportunities moved there in 1950 (first to nearby Sturdevant, Wisconsin) when she was 33 and he 39. They believed they could manage financially to educate their 11 children through the ninth grade. Just for them to be able to read and write or to obtain an office job which would not require the hard labor to which their parents were accustomed would be wonderful. Both level of aspiration and satisfaction with education for children scores were at the lowest level. Some of their children had married Anglos whose fathers were factory workers. Their friends spoke mostly Spanish with them and they had the highest possible Spanish usage score. Estralda's world view score was at the passive end of the scale and well formulated. Their possessions score was relatively high since they owned an old car and their home. However, the entire exterior of the home was in need of repair, the yard cluttered, the household in disorder, and the furniture scanty and in disrepair. Though her husband had held his job as an industrial laborer for the past 10 years, he did not work in 1958, at which time they were assisted by the Racine Social Services. In 1959 his industrial employment wages were supplemented with field work; they also received $600 from one of the sons, bringing the total family income to $3,500. In 1960 his hourly wages would give them a projected salary of $4,531.

Neither husband nor wife had any leisure time activities or hobbies, nor had either had any schooling. Her father was an agricultural laborer and she had worked in the fields since a child. Her husband, an orphan, had also been a field hand, although he had owned a farm for a period of eight years. Her antecedent handicap score was the highest possible.

Considering the above score, their age, the number of children, educational level, and language handicap, the chances of either Estralda or her husband being absorbed into the economy at high levels or integrated into the larger society were quite dim.

A Younger Negro Inmigrant More than
10 Years in Racine

By 1960, Elmer, who was 34, had lived in Racine for 13 years. His hometown was New Albany, Mississippi (1960 population 5,151), where he had also received his eighth grade education. Following military service he came to Racine on the advice of relatives who lived there. His father had been an industrial laborer and his father-in-law a craftsman. Elmer himself had held the same job as a craftsman since coming to Racine; in 1959 his total income had been $7,800. He wanted nothing less than a college education for his four children, though if a choice had to be made it was more important for the boy to attend college. Elmer's reason was: "To get a good job you gotta get an education." He picked such careers as nurse, teacher, doctor, and lawyer, any of which he thought he could manage financially. His possessions score placed him midpoint on the scale; the family did not have a fabric rug, a telephone, or a sewing machine, but had an old car, owned their home, and also had income from property at $35 per week. He and his wife attended parties, taverns, and movies; he belonged to the American Legion and liked pool, bowling, fishing, and/or hunting. Elmer's social participation score was in the high range. The exterior of the home and yard were in good shape but the interior was disorderly.

Elmer's world view was in next to highest active category; his level of aspiration as well as level of satisfaction with education for children were on the highest level. His antecedent handicap score placed him in the least handicapped half of the Negro sample.

Elmer's low antecedent handicap had apparently enabled him to overcome societal barriers and move upward in the economic order. His income was greater than most of the inmigrants who had come to Racine.

An Older Negro Inmigrant 10 Years in Racine

Perhaps one of the least successful examples of absorption into the economy is that of the following family:

Eleanor's husband had come to Racine in 1950 when he was 29. Previously he had lived in New Albany, Mississippi, where he had received six years of education. In 1960 he was a laborer (washed cars), a position he had held two years. Some of his prior jobs had been operative, but for the most part he had always been a laborer. His father had been a tenant farmer. Eleanor wanted their 13 children to receive a 12th grade education, but would be satisfied if they reached ninth. Her level of aspiration for the children and her level of satisfaction with education scores were the lowest possible. They had a close relative in Racine who worked as an operative, but they did not have any close friends; the occupational level of their associates score was in the lowest category. Her world view score was quite passive, and the family's possessions score was in the lowest bracket, though they did have a washing machine. Their income in 1959 was approximately $1,000 and it would not improve appreciably in 1960. They attended dances, parties, and taverns, while her husband went to movies and liked hunting, fishing, and baseball, putting both of them in a high social participation category. There was no yard, the entire exterior of the house was in disrepair, and the interior was disorderly with very scanty furniture. The hour-and-a-half interview was noisy, four adults and 11 children were in a small room with one couch, and the interviewer stood the entire time.

It is highly unlikely that the respondent (and her husband) will achieve a position of economic security. Antecedent handicaps, number of offspring, personal habits, and lack of education make occupational advancement difficult and a high level of living improbable.

A Younger Negro Inmigrant 10 Years in Racine

The last of the inmigrant life histories illustrates the case of a younger Negro who, although in Racine 10 years, had only been absorbed into the economy at a relatively low rate.

Olin had received his fifth grade education in New Albany, Mississippi, in the 1930's. He had worked as a taxi driver, section laborer, and coal loader. About 1950 he moved to Racine because he thought he could make more money and receive better treatment as a Negro. For 10 years he worked in a warehouse as a truck loader. His family received assistance from county welfare in 1960 and had numerous contacts with the Department of Social Services during their first years in Racine. His wife's part time job as a cook added an additional $500 to his wages so that in 1959 he had a total family income of $5,270. Olin and his wife had five children and though Negroes were treated a little better up North, they felt that 12 years of schooling was the maximum level that would benefit Negroes in securing a job. Since they did not believe in social relationships between black and white, their acquaintances were Negroes but none were close friends. They did not attend parties, taverns, or join clubs. Baseball and fishing were Olin's hobbies and his wife attended movies. What was unusual was the possession of two cars, one a newer than 1957 model being purchased on installment. Their rented house and the yard were neat and clean.

Olin's world view score indicated a definitely passive response. His antecedent handicap score was at the next to the highest level; his level of aspiration and level of satisfaction with education for children scores were in the lowest category. These scores are not unexpected considering Olin's fifth grade education, his low level first and present jobs, and his low income. That his possessions score was unusually high may be credited to his almost total non-participation in expensive leisure habits.

The diversity of these life histories does not lead one to expect antecedent sociological variables to be highly correlated with measures of economic absorption or cultural integration. Our purpose is to present a statistical picture of the chains of events which we have hypothesized as leading to economic absorption and cultural integration.

Correlational Analysis of Age and Time in the Community Groups

A Review of Groups Selected for Comparison

When the Mexican-American, Negro, and Anglo samples were partitioned according to length of time in the community—always lived in Racine, lived there 10 years or more, or lived there less than 10 years—and age 35 years or more or less than 35 years of age, there were sufficient cases to conduct separate correlational analyses for 14 race or ethnic, age, and time in the community combinations.

In all the age and time in the community categories, Anglos were at higher occupational levels than Negroes, and Negroes were at higher occupational levels than Mexican-Americans except for one case where all were at essentially the same level. The same relationship was found for income. The older Negroes who had lived in Racine 10 years or more were at higher occupation and income levels than any other category of Negroes; older Mexican-Americans who had lived there 10 years or more were slightly higher than other Mexican-Americans in terms of occupational level, but younger Mexican-Americans who had lived in Racine 10 years or more were better off in terms of income. When each of these age and time in the community groups were compared on level of living, Anglos, particularly older Anglos, had significantly higher scores in every instance than did Negroes or Mexican-Americans. Older Negroes and

Mexican-Americans who had lived in Racine 10 years or more had higher possessions scores than other Negroes and Mexican-Americans. In every case Mexican-Americans tended to have higher possessions scores than Negroes. In general, the Anglos, higher in every category, differed most from Mexican-Americans and Negroes on occupational level; Mexican-Americans and Negroes were most alike in terms of level of living, and differed to the greatest extent in income, with the Negroes higher than Mexican-Americans.

When age and time in the community groups were observed in terms of their level of education, Anglos were most educated, Negroes next, and Mexican-Americans by far at the lowest levels. Those Anglos, Negroes, and Mexican-Americans under 35 years of age who had lived in the community less than 10 years were best educated, while older persons in each time in the community group had less education. When level of aspiration for children was considered, all but one Anglo distribution was skewed towards high aspirations and that too was higher than any Negro or Mexican-American age and time in the community category. Negro aspirations were more similar to those of the Anglos than to the Mexican-Americans.

In summation, Mexican-Americans and Negroes who had lived 10 or more years in Racine and who were in the older age group were not always the highest on each characteristic within their racial or ethnic group. They tended to be: high on occupational level; high on income among the Negroes but low among the Mexican-Americans; high on level of living; lowest in education among the Mexican-Americans and almost as low among the Negroes; highest on level of aspiration among the Negroes and in the lowest category among the Mexican-Americans; and highest on level of satisfaction. More succinctly, older, longer-time resident (35 years of age and 10 years in Racine) Mexican-Americans and Negroes tended to have the status and that which Anglos consider to be indicative of success but did not possess the attributes by which Anglos would account for success in any Anglo society.

A Simple Four-Variable Model

As we approach the problem we are concerned with the relative correlation of one set of factors with another in some sort of meaningful time sequence, which we would describe as causal chains or experiential chains.[8] We shall commence with a simple model

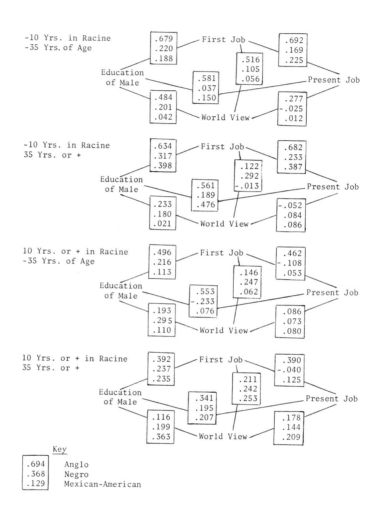

FIGURE 3. A FOUR-VARIABLE EXPERIENTIAL CHAIN

including only four variables: educational level of male, first job level, world view, and present job level.[9] The reader must remember that the correlation coefficients presented in Figure 3 tell us nothing about the relative standing of each race and ethnic group. What we have is a measure of the relationship of four variables to each other, one at a time, with controls for age of respondent and length of time in Racine.[10]

It is immediately notable that the two highest Anglo correlations in each age and time in the community group were between education of male and first job level, and between first job and present job level. The one exception was for those more than 10 years in Racine and under 35 years of age.

Secondly, each Anglo age and time in the community group (again with but one exception) had relatively low correlations between world view and education of male, first job, or present job. This time the one exception was for younger Anglos residing in the community *less* than 10 years. A third readily perceived fact is that the Negro and Mexican-American correlations were lower than those of the Anglos, with only four exceptions out of 48 possibilities. In half the cases, the Negro correlations were higher than Mexican-American correlations.

Most important, as we emphasized previously before holding age and time in the community constant, is the inescapable fact that educational level of male was correlated with first job for Anglos to a measurably greater extent than for Negroes or Mexican-Americans, with the highest correlation appearing for Anglos who had been in the community less than 10 years. We note that the correlations of first job level with present job level were consistently higher for Anglos than for Negroes and Mexican-Americans. This suggests that more job changes involving movement from one level to another have occurred between first and present jobs for the latter—largely an artifact of movement from agricultural labor to industrial labor or a kind of relative mobility characteristic of inmigrants from the South and Southwest.

That first job usually acts as an intervening variable (for all of the ethnic groups) is demonstrated by the higher correlations between education and first job than between education and present job. For Anglos, first job and present job were more highly correlated than education and present job (again with one exception).

The influence of world view on either first or present job is difficult to ascertain because of the possibility of conflicting

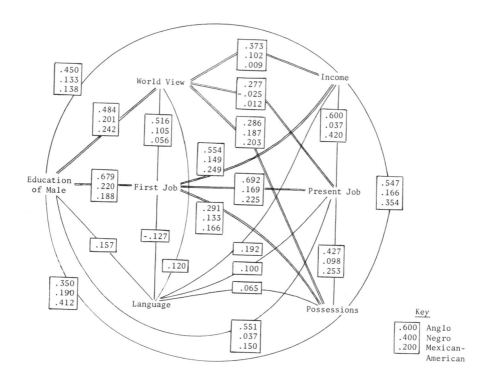

FIGURE 4. EXPERIENTIAL CHAINS OF MEXICAN-AMERICANS, NEGROES, AND ANGLOS LESS THAN 35 YEARS OF AGE AND WHO HAVE LIVED IN RACINE LESS THAN 10 YEARS

interpretations. Without longitudinal data it is difficult to say which is the antecedent. We can only affirm that world view was more highly correlated with education than with present job and more highly correlated with education than first job. That reciprocal interaction is likely should be recognized.

World view was most highly correlated with education, first or present jobs for Anglos less than 35 years of age and in the community less than 10 years, and next most highly correlated for Mexican-Americans older than 35 and in the community more than 10 years. Each had attained similar statuses in the community relative to their respective ethnic group but each achieved that standing from quite different origins and unequal advantages.

In conclusion, the hard fact is that education of male and first job are important determinants of present occupational level for Anglos. Not so for Negroes and Mexican-Americans. The path to occupational mobility is uncertain for any age and time in the community group among Negroes and Mexican-Americans. It is fortunate that first job is not highly correlated with present job—considering the level at which Negroes and Mexican-Americans enter the world of work. Furthermore, how one interprets the world about him seems to mean more if one is a young, newly arrived Anglo than it does for any other in our samples—and this may just as well be an artifact of his success as it is a factor in making it. It surely seems to have very little effect on the level at which Negroes and Mexican-Americans work.

This brings us back to an oft-repeated point—that the Anglos are more widely dispersed on each measure than are the Negroes or Mexican-Americans. The low coefficients of correlation for the latter two groups represent covariation at pretty much the lowest end of any measure. The coefficients presented in Figure 3 are merely further indication, for example, of how a little more education might result in a little better pay-off for some Negroes and some Mexican-Americans.

A Complex Experiential Chain

Let us turn to a more complex model of inmigrant experiences— the experiential chains of Mexican-Americans, Negroes, and Anglos less than 35 years of age and who have lived in Racine less than 10 years. In Figure 4 we again commence with education of male, probably obtained by Mexican-Americans in the Southwest, by

Negroes in the South, and by Anglos in Racine. We move to world view and first job for each of the racial samples but consider language just for Mexican-Americans. From these intervening variables we progress to income, present job, and possessions. Because of the complex interrelationships of correlations between education of male (which represents not only the amount of education but presumably reflects the influence of area of socialization), intervening variables, and income, present job, or possessions, we are also able to acquire some understanding of how the direct correlations between education of male and income, present job, or possessions are not as high as might be expected. As in the case of the simple four-variable model presented in Figure 3, it is not the correlations per se of each racial or ethnic sample which are so important as are the differences found when comparing the patterns of the three groups or the age and time in the community subcategories of each race or ethnic group.

We remain concerned with the relative strengths of different levels of determinants of economic absorption and cultural integration. Generally speaking, if antecedent sociological variables, such as place of origin and familial history, are of importance in determining the ultimate level of adjustment of each of the three groups in the study, they should be significantly correlated with the dependent variables—economic absorption and cultural integration. Likewise, if antecedent sociological variables are important determinants of the intervening variables (social psychological and otherwise) included in the study, they should be significantly correlated with them. And if the antecedent sociological variables have greater weight than the intervening social psychological variables in determining the ultimate level of adjustment of persons in the samples, they should be more highly correlated with adjustment measures. If the latter are more powerful determinants than the antecedent sociological variables, they should be more highly correlated with the dependent variables than the antecedents. It may also be that some intervening variables generated by the sociological antecedents create a cumulative effect. It is again possible that interpretation is not this simple—the social psychological characteristics of persons may be products rather than determinants of the extent to which they have been absorbed into the economy and integrated into the larger society.

We have selected younger inmigrants (less than 35 years of age) who have been in Racine less then 10 years for Figure 4, since of the four age and time in the community groups for which we have sufficient people to conduct analyses it is apparent that these Anglo respondents have a set of correlations most illustrative of that expected in our model of how society works for Anglos.[11] Figure 4 emphasizes, as in Figure 3, that the correlations between education of male and almost every other variable are considerably higher for the Anglos than for the Mexican-Americans or Negroes. What we also see is more pay-off income-wise and job-wise for Anglos than for Mexican-Americans and Negroes. Of all the variables intervening between education and income for Anglos, that with the highest correlation was first job (and this was the case for every age and time in the community group except that in which the respondents had lived 10 years or more in Racine and were less than 35 years of age, in which instance the highest correlation was between education and present job). Possessions are not as highly correlated with education of male for Anglos as for Mexican-Americans but the difference is small and might occur because there is more target buying on the part of inmigrant Mexican-Americans than Anglos or Negroes.

Moving to the influence of world view (an intervening variable) on measures of economic absorption, we again find that the Anglo correlations are higher than those of Mexican-Americans or Negroes. Here, of course, there is more difficulty in ascertaining with any degree of certainty that world view is a determinant of first job level, income, present job level, and possessions (an active world view is associated with high income and high level first and present jobs, and a high possessions score), or whether these measures have in turn been the determinants of world view. If we remember that first job is always highly correlated with world view (and generally more highly so than is present job) for all age and time in the community groups regardless of race or ethnicity, we can logically suggest that first job is more definitely related than present job to the formation of world view, and therefore that world view must have been formulated at an earlier period of time than a respondent's present job time—at least for a large proportion of our respondents.

While fewer years of education are related to greater use of the Spanish language for Mexican-Americans, the correlation is low not only in Figure 4 but for each age and time in the community group. Language usage, in turn, has a very low correlation with either present or first job, although the use of English is almost always associated with higher level first jobs and present jobs. In every age and time in the community group, English-speaking persons were more likely to have active world views. Those who spoke English most often had the highest incomes in every instance.

First job (Figure 4) is more highly correlated with income, present job, or with possessions for Anglos than for either Mexican-Americans or Negroes; this was generally the case regardless of age or time in the community. Anglos who started at higher levels for their first jobs continued to be at higher levels in subsequent ones, but there was no similar correlation between first and present jobs for Mexican-Americans and Negroes. They usually started at lower levels and remained there in subsequent jobs, though many had experienced the mobility that comes with movement from agricultural to industrial labor.

We have shown consistent differences between our samples of Anglos, Negroes, and Mexican-Americans while controlling for age and time in the community. Where correlations between antecedent or intervening variables and measures of economic absorption appear to be higher for Mexican-Americans than for Negroes it must be remembered that the relationships that exist for Mexican-Americans are based on scores falling at the lower end (below Negro scores) of almost any continuum that is selected.

The pattern of correlations for the group who, in 1960, had lived 10 years or more but not always in Racine and who were under 35 years of age was considerably different from the group in Figure 4. The Anglo correlations remained higher than those for Negroes and Mexican-Americans but the difference was not as great. A more important distinction, as suggested by the data for this group, is that there were other factors or variables for which we did not have measures that exerted more influence over present job level, income, and level of living for Anglos, Negroes, and Mexican-Americans than those we had measured and analyzed. The pattern of correlations that in part explained Anglo success as measured by Anglo criteria was not so distinctive as in the younger, shorter time in the community groups.

The third set of correlations, representing the interrelationships of variables for persons 35 years of age and over in 1960 and who had lived in Racine less than 10 years, showed Anglo level of adjustment varying with education and first job in the manner of the first group, but in this group Anglo first job had its highest correlation with present job. The level at which Negroes and Mexican-Americans are found is less likely to be related to their ascribed status as a Negro or Mexican-American.

The fourth set of correlations was based on the responses of those who had resided in Racine 10 years or more (but not always) and who were 35 years of age in 1960. Mexican-Americans, Anglos, and Negroes were more similar in many respects than in any of the other groups. Correlations between the antecedent sociological variables, the intervening variables, and the measures of adjustment were the lowest of the four age and time in the community categories. Anglo first job had its lowest correlation with present job; where they were in 1960 had less relationship to where they originally started and to what were supposedly the determinants of variation than in any of the other groups.

Social Background and Aspirations for Children from an Experiential Perspective

The next figure is somewhat different and commences with urbanization of home. First job, world view, and social participation are intervening variables between urbanization of home and present job, income, and possessions. The latter, in turn, precede (in time) level of aspiration for children and level of satisfaction with education for children.

Figure 5 presents the data for respondents who had lived less than 10 years in Racine but who were 35 years of age or over in 1960. This group was selected because the respondents were of the age most oriented toward children, they had been in the community a shorter period of time, and because we expected the differences between the Anglos, Negroes, and Mexican-Americans to be quite marked.

What do we find in Figure 5 that has not already appeared in Figure 4? Urbanization of home has very low or relatively low correlations with first job, with present job, with world view, with possessions, and with social participation of the males—with two exceptions. Its highest correlation is with first job for Negroes and with income for Anglos. First job is definitely correlated with present job, income, and possessions for Anglos and to a lesser extent for Negros and Mexican-Americans. World view has lower correlations with the same variables and with social participation of the male. Social participation of males has a relatively high correlation with present job level among the Anglos and Mexican-Americans but not among the Negroes while its correlation with income or with possessions is considerably less. [12]

Continuing across the figure, how are these variables related to level of educational aspiration for children and level of satisfaction

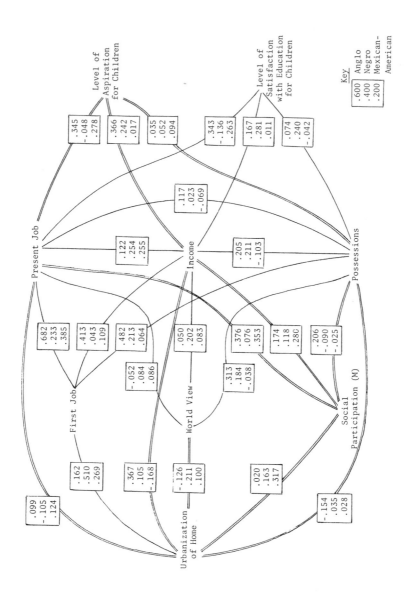

FIGURE 5. SOCIAL BACKGROUND AND ASPIRATIONS FOR CHILDREN FROM AN EXPERIENTIAL PERSPECTIVE, RESPONDENTS 35 YEARS OF AGE AND LESS THAN TEN YEARS IN RACINE

with education?[13] Present jobs and incomes are related to higher levels of aspiration for children, particularly for the Anglos. Similarly, high level present jobs are related to higher levels of satisfaction with education for the Anglos.[14] These correlations must be interpreted with care since, as it will be noted, the difference between Anglos, Negroes, and Mexican-Americans were not consistent from one age and time in the community group to another. Generally speaking, level of aspiration for children was high for Negroes and Mexican-Americans who were high on measures of socioeconomic status.

The complexity of the picture is emphasized, of course, by the fact that, while present job, income, and possessions have relatively low intercorrelations among Anglos in this group, they had relatively high correlations among them in the younger and less than 10 years in the community group.

There was little relationship between variables in the case of older inmigrants more than 10 years in Racine. Though some of the expected relationships, such as that between occupational level or income and level of aspiration for children or satisfaction with educational level, were found for Anglos and to a lesser extent for Negroes and Mexican-Americans, it cannot be said that the variables included in the model "explain" much of the variance in any of the samples.

The Influence of Associates

The next figure is just for Mexican-Americans and attempts to show the relationship of ethnicity of friends to intervening variables such as world view, language, and social participation, and the relationship of these intervening variables to present job, income, and possessions. The direct relationship of ethnicity of friends to present job, income, and possessions is shown at the opposite side of the figure. Mexican-Americans who had lived 10 years or more in Racine and who were less than 35 years of age were selected for the simple reason that the interrelationship of the variables was highest for this group. Those who had lived less than 10 years in Racine and were 35 years of age or more were least like the group selected. In both cases those who had Anglo as opposed to Mexican-American friends tended to have higher level present jobs, higher incomes, and more possessions. The fact that ethnicity of friends was highly correlated with present job does not enable us, however, to state that

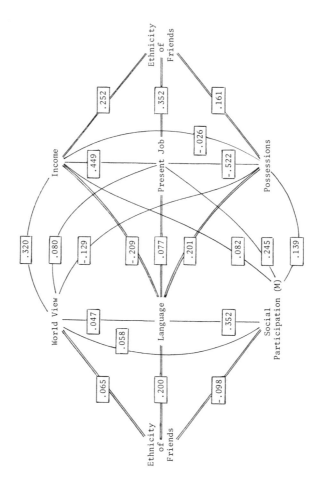

FIGURE 6. ASSOCIATIONAL CHAINS OF MEXICAN-AMERICANS IN RACINE 10 YEARS OR MORE AND LESS THAN 35 YEARS OF AGE

associations were the determinant of job level since it is just as likely (or more so) that the level of one's job can determine whether one has Anglo or Mexican-American friends. While ethnicity of friends among those in Figure 6 was correlated with language usage, its correlation with either world view or social participation was low.

Summary, Further Analysis, and Conclusion

Our major effort in earlier chapters had been to test a series of null hypotheses in order to determine the consistency of differences between race and ethnic groups and within segments of each race and ethnic group. The general hypothesis toward which we directed our analysis was that significant differences would appear between the Anglo, Negro, and Mexican-American samples on measures of economic absorption and to a more limited extent on cultural integration when controls were introduced for non-ethnic determinants. When those with the most of what makes for economic absorption in the larger society were compared, the Anglos were further removed from the Negroes and Mexican-Americans at the higher than at the lower levels. To put it bluntly, there was not quite so much concern over the race or ethnicity of a ditch-digger, but it helped to be an Anglo in the competition for an executive position!

However unwise it may have been considering the complexity of the data and the simple models that were presented, we sought to test hypotheses by comparison of Tau coefficients of correlation. Two difficulties were encountered: (1) most correlations were not statistically significant due to the size of the age and time subsamples on which they were based, and (2) two significant correlations (differing from each other in the order hypothesized) did not necessarily differ significantly from each other.

Our hypothesis of race and ethnic differences (in what are termed functional relationships) would be supported to the extent that Negroes and Mexican-Americans had lower correlations, or correlations in the opposite direction from those expected for Anglos, the Anglo pattern being that which would be present if the larger society operated as is generally believed.

Specifically, we hypothesized that early antecedents were more powerful determinants than intervening social psychological variables. Our hypothesis would be supported to the degree that correlations between antecedent sociological variables and measures of economic absorption were greater than the correlations between antecedent sociological variables and intervening social psychological variables (and between intervening variables and measures of economic absorption and cultural integration). Whatever the relationship, if the correlations accounted for only a small or relatively small proportion of the existing variance, then other more powerful variables at each time period or at each stage in the experiential chain must have been overlooked. There was also the question of the multiplicative effect of variables, or experiences in combination.

Most important, if only a small proportion of the variance in economic absorption between and within groups may be explained by the antecedent or the intervening variables that we have examined, we must again consider the impact of life experiences in Racine, as detailed in our associational hypotheses. How did the inmigrants relate to members of the host community so as to maximize or minimize the problems of absorption and integration? How receptive were their hosts? What had been done by the community to facilitate the processes of economic absorption and cultural integration? Here is the crux of the problem, for these are not a part of the simple models that were presented in this chapter.

When most major variables were correlated with each other for each year of the survey, one finding was consistent. Anglos, if not at the highest level on every measure, were able to translate what they had (their measurable marketable assets) into more of what was being measured than were similarly equipped Negroes and Mexican-Americans.

Approaching the data as a prediction problem resulted in essentially the same rank-order type of finding—Anglo status of one sort or another could be predicted somewhat more readily from a predictor variable than could that of Negroes or Mexican-Americans. Predictions on the combined sample were even more readily made.

We introduced a series of life histories to capture the trials in our respondents' lives—to represent the experiential chains that had brought them to their present economic and social position (1960). We demonstrated that the process of absorption and integration was most emphatically not the same for everyone, or at the very least, that given variables had different meanings and consequences as they appeared in this or that series of life experiences.

A simple four-variable model initiated the presentation of experiential chains. We chose one that would readily permit the reader to see how relationships varied with age and time in the community.

Figure 3, specifically with reference to the Anglo sample, clearly showed the intervening influence of first job level between education and present job. First job, although more highly associated with present job than education, was not so much higher that the critical role of education should be ignored. In fact, the high correlations between education level and first job demonstrated the power of education as a determinant of job level. World view, when conceptualized as an intervening variable, had little "influence" over present job level and was more highly correlated with education and first job. Since world view was not measured at such a time that it could be considered an antecedent to education and first job, it had to remain as either an intervening variable or a measure of cultural integration.

None of the correlations between education and first or present job was very high for either Mexican-Americans or Negroes as compared to those for Anglos. While in one-fourth of the cases the correlations between world view and educational level, first job, and present job were higher for Mexican-Americans and Negroes than for Anglos, the only clear finding for Negroes was that world view was more closely related to either education or first job than to present job and there was practically no relationship between world view and these variables for any of the Mexican-Americans except the older Mexican-Americans in the community 10 years or more.

The advantage of the simply constructed four-variable model was that it emphasized the sharp contrast between Anglos, Negroes, and Mexican-Americans, namely, the consequences of early educational advantages and early employment status on later employment status.

The second model, as shown in Figure 4 for respondents less than 35 years of age and who had lived in the community less than 10 years, was, in essence, an expansion of the four-variable model with the addition of language, income, and possessions. The complexity of the model became far greater with the addition of variables. Present job, income, and possessions are more closely interrelated among the Anglos in only this age and time in the community group—in others there was no definite pattern. Nonetheless, education of male and first job were generally more highly correlated with measures of absorption for Anglos than for Negroes or Mexican-Americans. For the Mexican-Americans, usage of the English language was of less importance in determining level of absorption (except for income but the difference was small) or cultural integration than was level of education. On balance, one must conclude that education, as mediated by first job, was the most powerful determinant of level of economic absorption for Anglos and the most important for Negroes and Mexican-Americans among all the variables presented in the model. But in the latter case, the path to absorption remained unclear for we did not include in this model other variables that correlate with economic absorption for Negroes and Mexican-Americans.

The third model was the most complex. Here we attempted to show how the basic variables included in the two previous models were correlated with other indicators of life experiences and later measures of cultural integration as well. We selected a different age and time in the community group for Figure 5. The influence of urbanization of home turned out to be small—this should not be surprising to the reader since various measures of urbanization were mentioned in earlier chapters as having little relationship to other variables included in the study.

On the contrary, level of educational aspiration for children and level of satisfaction with education for children, two similar measures, were correlated with present job and income and usually to a greater degree for Anglos than for Mexican-Americans and Negroes, but in several instances more highly correlated with income and occupational level of Negroes or Mexican-Americans than for others. This suggested that level of aspiration as an indicator of cultural integration may be more closely related to the inmigrants' socioeconomic status than to his race or ethnicity. The meaning of the correlation of social participation of males with occupational level and income was difficult to determine since it cannot be said that either is an antecedent of the other. If it may be assumed that participation in social groups is an antecedent then it could certainly be said that it had a varied but most notable impact on older persons less than 10 years in the community.

The fourth model, as shown in Figure 6, concerned only the Mexican-American sample. Of what influence were Anglo or Mexican-American friends? Ethnicity of friends had relatively low correlations with each age and time in the community group; it made no difference whether it was considered an antecedent to world view, language usage, and social participation, or as an antecedent to present job, income, and possessions.

What conclusions can we draw from the above intricate figures and explanations? Most certainly we have not "explained" the variance in level of absorption among Anglos, Negroes, and Mexican-Americans, either taken as a group or as compared within groups. But certainly, and about this we need not hesitate, the data showed that negative restraints were responsible for group differences in two ways: (1) in determining the level and type of education acquired and initial work opportunities for Negroes and Mexican-Americans at point of socialization, and to an extent in the host community, and (2) in preventing Negroes and Mexican-Americans from advancing in the economic system as measured by present job and income on a basis of the capacities that they acquired (despite the restraints placed upon them). Whatever progress they had made was to some extent limited by the manner in which gate-keepers, both public and private, have operated at various levels in Racine.

The consequence of individual differences in education, experience, and level of aspiration is the generation of differences in occupation and income that are more readily visible and measurable for Anglos than for Negros and Mexican-Americans, at least in 1960.

What we have is a statistical picture of the confusion and contradictions in the larger society—a type of empirical evidence of how life's experiences are frustrating for some Anglos, of course, but far more so for urban Negroes and Mexican-Americans.

NOTES

1. See Elaine M. Krass, Claire Peterson, Lyle Shannon, "Differential Association, Cultural Integration, and Economic Absorption among Mexican-Americans and Negroes in a Northern Industrial Community," Southwestern Social Science Quarterly, December, 1966, pp. 239-252.

2. See S. M. Miller and Frank Riessman, "The Working Class Subculture: A New View," Social Problems, Vol. IX, Summer, 1961, pp. 86-97; A. B. Hollingshead and F. C. Redlich, Social Class and Mental Illness, New York: John Wiley & Sons, 1958; Robert A. Nisbet, The Quest for Community, New York: Oxford University Press, 1953; Gresham M. Sykes and David Matza, "Techniques of Neutralization: A Theory of Delinquency," American Sociological Review, Vol. XXII, December, 1957, pp. 664-670.

3. See Krass, Peterson, and Shannon, op. cit., Table 4, p. 250.

4. The inmigrant southern white is often in much the same circumstances, as several studies have indicated, in that he too is slow to become similar to well established residents of the community. See for example, Grace Leybourne, "Urban Adjustment of Migrants from the Southern Appalachian Plateaus," Social Forces, Vol. 16, December, 1937, pp. 238-246; Morris G. Caldwell, "The Adjustment of Mountain Families in an Urban Environment," Social Forces, Vol. 16, March, 1938, pp. 389-395; Lewis M. Killian, "The Adjustment of Southern White Migrants to a Northern Urban Area," Social Forces, Vol. 32, October, 1953, pp. 66-69.

5. See Lyle Shannon and Patricia Morgan, "The Prediction of Economic Absorption and Cultural Integration among Mexican-Americans, Negroes, and Anglos in a Northern Industrial Community," Human Organization, Vol. 25, No. 2, Summer, 1966, pp. 154-162.

6. In order to combine variables, several item analysis scales were constructed. Each item in the scale had a weight proportional to its correlation with the variable to be predicted. Separate scales were constructed for Mexican-Americans, Anglos, or Negroes and one scale from a combined sample of the three ethnic groups. Occupational level was predicted from an item analysis scale based on education of male, occupational level of first job, occupational level of associates, and extent of social participation. World view was predicted from a scale based on male education, present occupational level, occupational level of first job, occupational mobility of male, occupational level of associates, and social participation.

7. See Shannon and Morgan, op. cit., Tables 4A and 4B, p. 162.

8. There is a very extensive literature on causal chains and path analysis. See, for example, Hubert M. Blalock, Jr., *Causal Inferences in Nonexperimental Research,* Chapel Hill: The University of North Carolina Press, 1961, 200 pp.; Otis Dudley Duncan, David L. Featherman, and Beverly Duncan, *Socioeconomic Background and Occupational Achievement: Extensions of a Basic Model,* Final Report on Project No. 5-0074 (EO-191) for U.S. Department of Health, Education, and Welfare, Office of Education, May, 1968, 296 pp.; H. M. Blalock, "Four-Variable Causal Models and Partial Correlations," The American Journal of Sociology, Vol. 68, 1962, pp. 182-194.

9. For a more sophisticated model using path coefficients for the antecedents of educational and occupational attainment, see William H. Sewell, Archibald O. Haller and Alejandro Portes, "The Educational and Early Occupational Attainment Process," American Sociological Review, Vol. 34, No. 1, February, 1969, pp. 82-92; William H. Sewell, Archibald O. Haller and George W. Ohlendorf, "The Educational and Early Occupational Status Attainment Process: Replication and Revision," American Sociological Review, Vol. 35, No. 6, December, 1970, pp. 1014-1027. The models proposed by Sewell, Haller, Portes, and Ohlendorf relate educational and occupational attainment to level of occupational aspiration, level of educational aspiration, significant others influence, academic performance, socioeconomic status, and mental ability. They have applied them to young men in Wisconsin; their adequacy for very large cities (Milwaukee is the largest in the sample) has not been established but the authors doubt that the results would be greatly different in great metropolitan areas.

10. In earlier chapters considerable attention was paid to the significance of differences between distributions for the Anglos, Negroes, and Mexican-Americans and the significance of differences between subcategories of these samples. We also tested a variety of null hypotheses, hypotheses based on a model of no relationship of one variable to another within a race or ethnic category or subcategory. That significant differences exist has been well established in these earlier chapters. At this point we are mainly concerned with measures of association in order to make comparisons within and between race and ethnic groups. Some of the low correlations are not statistically significant and in some

cases it would not be proper to assume that the differences in correlations between groups and subgroups are great enough that other samples from the same population would be similarly rank-ordered (in terms of correlation coefficients) in exactly the same way. However, it is safe to say that these patterns of correlation differences do demonstrate the variability of the interrelation of variables to which we refer in this chapter.

11. Unfortunately, there were relatively few Anglos (16) in this group, although relatively large numbers of Negroes (76) and Mexican-Americans (43).

12. We should note that social participation of the female has its highest correlation with social participation of the male—the correlation for Anglos being highest, followed by that for Negroes, and lowest for Mexican-Americans.

13. The highest correlations between social participation and level of satisfaction with education for children, and level of aspiration for children are for male Mexican-Americans. Although these correlations are still relatively low, they do suggest that participation in the activities of the larger society does have an influence on how the Mexican-American inmigrant perceives the future for his or her children.

14. One further relationship that was explored but not included on the figure was that between tavern attendance and level of aspiration for children and level of satisfaction with education for children. Anglo males who attended taverns were a bit more likely to have a high level of aspiration for children while the opposite was true for Mexican-Americans and Negroes. The differences (Anglo .233 and .257, Negro −.247 and −.213, and Mexican-American −.257 and −.153), while rather intriguing for this age and time in the community group, are not present in any other age and time in the community group. In fact, no two groups have the same pattern.

Chapter 13

IMPLICATIONS OF THE STUDY FOR

COMMUNITY ACTION

In a final report to the National Institutes of Health, our basic source of the supportive funds for this research, considerable attention was devoted to the implications of our findings for public health medicine and social welfare programs.[1] The scope of this volume permits us to apply the findings not only to the delivery of medical care and welfare services but to a broader spectrum of problems, particularly to those in the educational institution and the world of work.[2]

Problems for Inmigrants and Minorities in the Educational Institution

The importance of Negro and Mexican-American perceptions of the value of education cannot be overstressed, especially as portrayed in their responses pertaining to educational aspirations for children and in those where education was mentioned as a vehicle to better jobs and upward mobility, or in their words, "getting ahead." Though many gave obviously unrealistic responses in terms of the occupational goals they had specified for their children, they were convinced that "education" was a necessary (if not sufficient) prerequisite to their attainment. Herein lies a paradox. Education must bring about enough intergenerational change to be identifiable as a crucial factor in changing people's lives. Children must recognize that the efforts of their parents to keep them in school have "paid off" so to speak; parents must also be able to observe the tangible benefits of education for their children. But if the "pay-off" for increasing increments of education is not the same for Mexican-Americans and Negroes as it is for Anglos, then disillusionment is almost certain to be generated.

Consequently, we must be concerned about the quality of education and to whom it is available rather than simply the number of years of schooling.[3] The community should recognize that opportunities for higher education are unevenly distributed in the population, particularly in reference to race and ethnic backgrounds.[4] Proper (and unbiased) counseling for the changing and increasingly complex world of work should be of paramount concern.[5] If fewer and fewer positions in the system of production require non-technical skills, then how may this relatively uneducated population of adult Mexican-Americans and Negroes be channelled into the kinds of educational opportunities that will open the door to their absorption into the economy?[6] Should an unemployed foundry worker who seeks new skills to beat his joblessness be denied retraining?[7]

Adult Education

The role of education as it relates to the absorption of inmigrants into the economy may be thought of as beginning with the parents who have come to Racine, bringing with them their children, or presenting them to Racine at a later date, and concluding with the cultural integration of these children. From this viewpoint it seems logical for a community such as Racine, which needs proportionately fewer and fewer unskilled workers, to commence with the task of improving the educational opportunities of the adults. But is the retraining of workers defined by any community as its responsibility? Is it rather the joint responsibility of (1) industries that have displaced workers, (2) industries that need trained workers, and (3) the citizens of the community through the public school system?[8] This is the present dilemma of most northern industrial communities, a problem not dissimilar to that of southern and southwestern communities once employing many thousands of agricultural workers. There is also the question of the relative importance of simple literacy training, elementary and secondary education, and on-the-job training. Although the question of which elements of the community are most responsible for adult education is controversial, there is agreement that previous efforts have not really been for those at the lowest levels of the economy.[9] They have always had the least of what is needed to cope with changes in the organization of production.

Education of the Youth

The problems of elementary and secondary education in reference to inmigrant children are multiple. Retention in school, communication while in school, return to school after drop-out, and continuation beyond secondary school to technical school, college, or the university are some of the basic concerns, though not exclusively the problems of the inmigrant.

Successful communication with minority group and lower socio-economic status students, especially in inner-city schools, cannot be overemphasized. Before the teacher can expect much success in the classroom, various language problems must be taken into account in designing the curriculum. Gloria Channon has divided the problem into four major aspects: the phonological, the lexical, the syntaxical, and that of attitude.[10] She takes the position that from the beginning the teacher must be aware of differences in standard and non-standard speech. Secondly, through pictures, materials, films, tapes, and conversations, the child must be assisted in labelling *his* world—a strategy at variance with the one which preemptorily labels what is for all practical purposes a non-existent world for the child. Thirdly, the child must be given practice in hearing and seeing his thoughts expressed, not only in his words but in standard English. Last, and most difficult of all according to Channon, the attitude of teachers must be changed. They should, for instance, discard their outdated ideas about the positive relationship of verbal facility to innate intelligence.

Pertinent to the above statement is Arthur B. Shostak's contention that testing programs should be re-examined because deprived children have demonstrated considerable intelligence in the way that they cope with their bleak environment. Less use should be made of standard achievement or aptitude tests and in their place should be substituted new "culture-fair" tests, currently under development, that enable one to predict performance in real life situations. His idea that poor children should be given practice in test-taking is cogent, for if a low score is due only to lack of experience in testing situations, then indeed they should be given the "practice" that will permit performance at a level which better reveals their ability.[11]

Shostak's concern over the consequences of testing programs would not be so significant were it not for the fact that school teachers, and of course their administrators, believe that testing plays such a crucial part in placing the child in relation to his peers in the classroom—the entire process supposedly being for the benefit of the child rather than for the convenience of teachers and administrators. Beyond the perpetual disadvantage that may accrue to a child as a consequence of placement which minimizes interaction with future competitors is the effect of teacher expectations on his achievements.[12]

So much has been written (and published) about the importance of cultural differences as determinants of behavior that little can be added except to reiterate that teachers in an industrial community should not expect all of their students to have backgrounds similar to each other or to that of the teacher. Since the "middle class" teacher has not always appreciated socioeconomic and religious differences among students in the classroom, he is all the more unlikely to perceive and understand the differences in the antecedents of Mexican-Americans and Negroes. To what extent special institutes for teachers might assist in solving this problem is uncertain. Since they do accept institutes, emphasis on the changing composition of the community might be an appropriate concern for the training of both public and parochial teachers.[13]

Offenbacher points out that two sets of policy recommendations for the educational system have evolved: (1) divesting the school of some of its middle-class biases, and (2) enriching the culturally deprived background of the lowest class child. According to the findings of her study, the school cannot afford to follow the first and the majority of the lower class students have little use for the second. She concludes that the answer may be to ask the culturally privileged to share their wealth with the culturally deprived.[14] One suggestion has been to establish a one-to-one relationship between each culturally underprivileged child and a non-professional buddy. (Here the question may be who comprises the culturally deprived? At one point it was fashionable to describe the middle class children as such. A university president recently stated that middle class students in universities were "deprived" because their early years were devoid of hardships and thus they had missed the opportunities of depression-reared youth.)

The possibility of bilingual education, at least in the earlier years, has been considered by a number of groups. We refer the interested reader to "The Invisible Minority," Report of the NEA-Tucson Survey on the Teaching of Spanish to the Spanish Speaking.[15] The value of such a program, of course, is different in Racine than in the Southwest where bilingualism is almost a prerequisite to successful interaction with members of the larger Anglo society.[16]

Admittedly there are problems outside of school jurisdiction which tend to complicate the communication system between parent, student, and the teacher. Schools have no control over the high mobility among migrants who find it difficult to become absorbed into the urban economy, or mobility related to visits with relatives in the Southwest. To insure the continuity of education, teachers and administrators should fully understand the need for formal and informal networks of communication reaching both children and their parents. The administrator who told us at our first meeting with him that he didn't have Mexican-American parents in the P.T.A. but would have some as members by the next meeting indicated his recognition of a basis for interchange—but failed to realize that extensive interaction between members would not be possible unless the language handicap were overcome.

Beyond the problem of retention of the student in school is the one of bringing students back into the school who have dropped out. Here the community needs to fully understand the nature of the problem and to give encouragement to administrators or teachers. Working with students who have not been "fired up" sufficiently by the educational process to remain in school is not an easy task. It may be that the usual academic program and atmosphere of quiet and orderliness so valued by "middle class" teachers is not conducive to the success of a program for Mexican-Americans and Negroes who are already considerably older than their more fortunate Anglo classmates.

Establishing store-front schools for drop-outs is a current attempt with possibilities. The Urban League has established over 30 of these schools in New York City, some of which are job training centers and others of a pre-academic nature to prepare students for advanced educational opportunities.[17]

If the school were perceived as a "center" of the community rather than as an "institution" which young people, at least in the eyes of some, are forced to attend by representatives of the larger community, the problems of adult education and drop-outs might not be as difficult. Henry Saltzman and others have suggested that schools should be operated as community centers on a seven-day week, 12-months-of-the-year basis. [18] With the extensive use of busing in some urban communities it might be difficult for young people from non-contiguous areas to have their school as a community center, but any school can at least become the center for adults and perhaps an activities center for people of all ages in the area in which it is located.

Higher Education for Minority Groups

No matter how highly parents value education, their children do not automatically perceive it similarly. And sometimes when they eventually do, their high school performance has been such that it is too late to aspire to college or the university. Sometimes the immediate reward for those who do pursue higher education is difficulty with their peers. For example, an athletic scholarship has been defined by Negroes as a legitimate method of reaching a legitimate goal (college), while an academic scholarship has been perceived as an illegitimate means of achieving the same. [19] Again, the Negro who achieves high grades in school may also be labelled an "Uncle Tom" by his peers. Petroni found Negroes who attempted to achieve academically to be labelled more quickly by their peers than were Mexican-Americans who tried to compete with Anglos in intellectual, political, and cultural activities. While this reaction is understandable as a movement away from the "oppressor," it nevertheless indicates the price that some must pay while engaging in behavior which will facilitate their upward mobility.

In recent years some educators and non-educators who have become interested in increasing the ability of minority groups to compete in the larger society have pushed for various programs and policy changes that would facilitate entry into institutions of higher learning. While the community may decide that an attempt should be made to increase the proportion of young, able Mexican-Americans and Negroes who seek entry into colleges, universities, or technical institutes, concentration on increasing the number of applicants, regardless of readiness and chances of securing admission, might not be as effective as providing financial assistance to those who are able to meet the rigorous entrance requirements. Which brings us to the question: who will benefit from education or increasing increments of education? In a recent article in Science, [20] Julian C. Stanley discusses the damage that can result from placing young disadvantaged people in college simply because they might benefit from the experience. The problem is not solved, he decides, by pushing unqualified minority groups into training which they have little likelihood of completing.[21] What must be done is to place persons in the kind of training at which they are most likely to succeed and which will enable them to increase the chances of reaching their goals. Stanley suggests that instead of being *less* concerned about the qualifications of the so-called disadvantaged applicants, the trend should be reversed and we should be *more* concerned about them. If we follow this premise we may find that our present primary and secondary programs have not produced as many qualified Mexican-Americans and Negroes as we believe should be in colleges and universities.[22] One wonders if Mexican-Americans and Negroes have been proportionately included in all contemporary attempts to rectify disadvantages—the Headstart Program and more recently Home Start, for examples.[23]

Admittedly what has been presented on the educational institution is incomplete in terms of the countless problems, fiscal and otherwise, encountered by administrators and teachers. But the underlying basic concern still remains. How adequately has this specific nucleated institution adapted itself to the needs of an inmigrant people and correlated its efforts with other social institutions in the community to cope with the changing composition of the population and changing modes of production?[24]

The World of Work and Welfare Economics

As early as the pretest period in 1959 it was apparent that most Mexican-Americans had moved to the North in order to secure employment; the same was true for each of the inmigrant groups of Mexican-Americans and Negroes interviewed in Racine. In the pretest city where a Mexican-American community had been well established, very few persons indicated any intention of returning to their previous place of residence. In fact, half the respondents in the pretest community mentioned relatives or friends who wished to come to the North to better their economic situation. Mexican-Americans and Negroes in Racine reacted in the same fashion, though to a somewhat lesser extent.

But however much the decision to move is prompted by economic or work-oriented reasons, there has always been a very imperfect process of fitting the available labor force to the work opportunities present in any community.[25] In even a relatively small urban industrial setting, such as Racine, no one knows how many jobs or what types of positions are available from year to year, much less from month to month or week to week. Nobody knows the qualifications, if any, and the demographic characteristics of persons who fill these positions, or of those who have not secured positions but are searching for a place in the economic order. Simply stated, there is little knowledge of the human dimensions of the economy in Racine from week to week and month to month. And if someone were to request detailed information in reference to cities to the north or south of Racine as a basis for advising job-seekers it too would be unobtainable.[26] The situation in Racine today (1973) is just as it was in 1960. That this basic statistical information is completely lacking or available only in partial and more or less inaccurate form works more hardship on those who need the information most, namely the relatively unstable, unintegrated group who are in transition from farm labor and migratory work status to that of urban industrial laborers.

We have taken the position that within the so-called "lower class" culture there are subcultures. One includes unskilled irregular workers, frequently inmigrants to the urban industrial setting, who lack the economic security needed to cope with their new way of life. This subcultural group differs from another which consists of more stable *working* class members who have already successfully attained a solid foot-hold in the economic institution and who are ready to acquire a more middle class orientation.[27] Among the first group, the hard core, hard-to-reach, or chronically dependent families, the importance of economic absorption would seem to be an antecedent to complete cultural integration.[28] Levinson, in his conceptualization of chronic dependency, has termed the involvement of the hard-to-reach client in treatment as alienative. He would probably categorize the inmigrants in this study as alienated from middle class social institutions because they lack any real social involvement with the community outside of their family and place of employment—if they have any.[29]

Hanson and Simmons offer an introspective answer:

"For most rural migrants to the city, the key to the adjustment process is stable employment which produces a steady income which must then be managed properly. But the typical public or private employment agency provides little help for the migrant needing *stable* employment. Perhaps a working answer is some partial intervention into the network of informal social relationships the migrant soon encounters as he develops friendships and acquaintances. A 'gatekeeper structure', consisting of recognized ethnic group leaders, could be organized to provide information to incoming migrants. Ideally, personal friendship or acquaintance linkages would connect the leaders who have access to reliable and sound information at the top, with the migrants newly arrived in a neighborhood or job. Then the experience path of the newcomer would be monitored by a friend or acquaintance tied into the gatekeeper structure. When misfortunes arise, or a job is lost, the migrant would have access, through his friend, to the person or persons at higher levels of the structure who have the appropriate knowledge for handling the problem."

Suggestions for Public Health and Social Welfare

At the inception of this study, some public health and social welfare workers had expressed concern about their failure to reach what are often called unsocial, hard-to-reach clients. Casework techniques seemed to fail in bringing about any real changes in behavior. Clients of this type were often either brought involuntarily to an agency by a court, or the agency unsuccessfully expended its resources in "reaching out" to them.[30]

There are two ways of approaching the problem: we may choose to aid the inmigrant in the ways he perceives as being most effective in achieving absorption and integration or we may concentrate on communicating with the inmigrant as he progresses through the formal institutional channels established in the community. In order to transmit public health practices and modern medical knowledge most effectively to inmigrants as suggested in the first approach we must relate these services to the changes that the inmigrants have been seeing in their move from former places of residence, the changes that have already taken place in their lives, and how they perceive what has happened as a consequence of their move. In respect to the second approach, we must carefully examine the position of the inmigrant in the economy and the process by which absorption takes place in order to successfully introduce new or modified values or behavior patterns.

In an urban community the heavy industries such as mills, foundries, and forges are most likely to employ the inmigrant worker. A study of the ecological distribution of Mexican-American workers in a steel mill in the pretest community in 1959 revealed that some departments had few or no Mexican-Americans and that other departments employed anywhere from 50 to 80 percent. Most of the major decisions of the inmigrant are ordinarily made within an economic framework or within the realm of his world of work. He is also most likely to learn or acquire knowledge of public health practices or modern medicine in his daily working environment because many industrial plants insist upon a physical examination as one of the first requisites for employment. The industrial worker may also be given periodic check-ups; if he is injured on the job he is sent to the dispensary; he can become a member of a group medical plan. An excellent maneuver by concerned agencies to facilitate the transmission of additional public health or medical knowledge to the inmigrant male worker within a work context would be to set up easily accessible offices within the concentrated areas of the industrial plant in which the target population works. Cooperation between the community's public health and social welfare services and these industries could show as a next step how pertinent family health is to the health of the male wage earner. Community health and welfare centers that are decentralized and open in the evenings, rather than located in central congested business areas with hours eight to five, would make it far easier for families to make contact when in need of assistance.[31]

Prestigious persons from the Mexican-American and Negro communities acquainted with public health and welfare agencies could be selected from within the industrial plant or other work situations in order to assist in the referral process at either industrial or community health centers. Here we are again suggesting that the informal patterns of association developing from existing work groups serve as the best channels of communication among those who are being absorbed into the economic institution. Knowledge of the informal status structure of the inmigrant group is crucial to successful implementation of this approach.

The question is, of course, what to do about those who are not being absorbed into the industrial economy or those who have only the most tenuous relationship with the urban economy. These persons are less likely to visit the dispensary, undergo medical examinations, or participate in company health programs. The situation of the continually unemployed inmigrant is, of course, obviously worse since these sources of medical knowledge are completely lost to him. In such cases, the educational institution may be correct institutional source for transmission of information about public health, medicine, and social welfare service. All inmigrants, regardless of work or non-work status, tend to place great value upon education, though many may be relatively ignorant of how or how much education facilitates absorption into the economy.

Perhaps we should turn again to the idea of the school as a multipurpose community center. If the physical health of children is a prerequisite to ability to learn, it would seem good strategy for the community to bring health and educational institutions into a close physical and working relationship. The current practice of employing school nurses cannot be considered the same as intermeshing the educational process with the process of achieving better delivery of health information and medical care to the whole inmigrant family, let alone the community. If the adult inmigrants as well as long-time residents in the community define the school as a center for them and for their children, then what the school, with the cooperation of other institutions has to offer is more likely to be sought after and accepted. This is more probable if the center is perceived responsive to the needs of the community rather than as a creature of outsiders who have more to gain from the success of the program than the intended recipients.

Public health nurses have in the past found it difficult to establish rapport with inmigrant families, suggesting that the inmigrants defined them as persons who could not facilitate the acquisition of anything the Mexican-Americans and Negroes sought as a consequence of moving from the South and Southwest. One barrier was language—Mexican-Americans generally know only Spanish (or prefer to speak Spanish), which places the Anglo, English-speaking, uniformed public nurse or similar representatives of other Anglo institutions at a disadvantage in communicating with inmigrant workers. Despite the fact that the professional social worker is trained in methods of establishing rapport with an individual client, the very nature of the structure of a social agency, its formal organization, and its tendency to "check" on people, present additional obstructions to intimate contact.

Compounding the problem is the difference in socioeconomic status or "social class background" between most public health and welfare workers and their inmigrant clients. It has been difficult for the traditional social worker, with a value orientation differing from that of his client, to assist him according to his (the inmigrant's) values and to help him attain only those changes in his life that he distinctly desires. It has been almost impossible, at least until the sixties, for social workers to conceptualize any of several "lower class" subcultures as social systems with characteristic sets of practices, focal concerns, and ways of behaving that are meaningful and systematically related to one another. They fail to understand that lower class inmigrants perceive their move to the northern industrial city as at least a partial fulfillment of their aspirations and a significant change in their position. They term these inmigrants as troubled or disturbed because they are not engaged in the kinds of activities defined by themselves as likely to be productive of further mobility. They do not see the inmigrant as a member of a relatively stable ethnic or racial lower class subculture with somewhat different standards.[32] If social workers, paraprofessionals or other professionals for that matter, could accept the concept of cultural relativity—a concept once deemed of paramount importance by sociologists—then it would be only one step further for them to adopt the objective of assisting people to adjust to the new environment effectively within their own social class or subcultural tradition.[33]

The need for less formal patterns of contact between the personnel of agencies and inmigrants might be accomplished by establishing a detached worker program which includes intermediary professional or paraprofessionals who aid the social worker in the community rather than in the office. Such a person, preferably female, could reach the women who seem more often less assimilated than their economically absorbed husbands.[34] Carefully choosing personnel for this type of social welfare program would maximize identification of the client with the worker. It at least offers more diffuse, community centered, and informal methods of reaching and treating the industrial inmigrant than does the typically urban industrial type of community organization with highly specialized, bureaucratic, and centrally located social welfare agencies. Other organizations, religious, political, recreational, or social with a potential for more informal contact with the inmigrant population than the currently structured social welfare agency, could also develop a social service framework for reaching the Mexican-American and Negro population.[35]

In addition to structural revision of the present social welfare system, the methodology of casework practices has long needed revision.[36] A "cultural"[37] frame of reference which seems to be a more appropriate way of explaining much of the behavior of inmigrants ought to be substituted for the individual psychological approach or psychodynamic concept of social casework. Since a personality is shaped largely by the culture in which socialization has taken place, an understanding of the cultural traits of various ethnic or other groups would facilitate casework.[38]

As the inmigrant population recognizes that changes are being made for the better, its level of aspiration will rise. Although the ecological segregation of the inmigrants tends to reduce opportunities for comparison of their way of life to that of the majority population of the community, there is sufficient awareness of the life style and reward system available to most Anglos to produce an element of dissatisfaction. The aspirations of inmigrants may be expected to increase proportionately to this awareness and further so as their perceptual horizons are broadened by more contact with Anglos.

Communication Through Voluntary Organizations

The literature suggests that Mexican-Americans do not join as many organizations, particularly voluntary ones, as do Anglos. This has been corroborated by our findings in Racine. It is quite possible that these organizations do provide a mechanism for communication between and within ethnic and racial groups that should not be overlooked. It is true, of course, that most organizations are not truly inter-ethnic or inter-racial (though they no longer admit to being discriminatory). Most Anglos are also totally unfamiliar with the organizational structure of either the Mexican-American or the Negro community. Negroes in Racine, for example, choose the larger formal organizations, such as the NAACP or the Community Action Program and its numerous neighborhood centers as points of involvement. It is rather clear that the LULAC's and the Spanish Centers would reach the largest numbers of Mexican-Americans. Nor ought the importance of churches in the process of communication be forgotten. Mention of the role of voluntary associations, however, does not reduce the importance of community or particularly the Mexican-American community where the heads of key families are important starting points for dissemination of information.[39]

Conclusion

While many have spoken and written in recent years about the cost of social welfare programs, an equal number have taken the position that no one should suffer from want in a wealthy, highly productive society. This chapter is based on the assumption, as Ornati puts it, "that one of the functions of modern cities is to transform the poor in-migrant into the non-poor participant in urban life—and the meaning here is modern life."[40]

Those in control have traditionally utilized their power to perpetuate the kind of social organization that they believe to have been most effective in reaching the goals they value. More recently the less fortunate, first Negro and later Mexican-American, have utilized a variety of strategies in attempts to modify the organization of society and increase their life chances.[41] The community has often been reluctant to accept some of the techniques that have been utilized by the less fortunate. At the same time, realizing that these groups have found it difficult to accept their position of powerlessness, Community Action Programs were initiated in urban areas and to a lesser extent in rural areas. These were designed to increase the participation of the "poor" in programs that would benefit them.[42]

This research indicates that Mexican-American and Negro inmigrants to Racine make up a disproportionate share of the disadvantaged, of the less fortunate members of the community. If they engage in behavior which they believe will increase their life chances or assist them in reaching their goals, it should be expected and interpreted as normal behavior in a society organized in the fashion of our urban industrial societies.

Henry Miller's story of a guru,[43] an Indian wise man who sat with a cocoon in his hand, serves as a fitting conclusion to this chapter:

> "He [the guru] was watching the cocoon when a little boy came to him and asked him, 'What is that?' and the guru said to the little boy, 'It is a cocoon and inside is a butterfly. When the time comes for the butterfly to come out, he will break open the cocoon and come out to fly by himself.' The little boy then asked the guru if he could have the cocoon. The guru said, 'Yes, you can have the cocoon, but you must promise me, little boy, that when the cocoon splits and the butterfly starts to come out, no matter how much you want to help the butterfly you mustn't do it.' The little boy agreed to this promise. He took the cocoon home and he sat and watched and finally in a little while the cocoon cracked open. Inside was a wet, moist, butterfly beating against

the shell of the cocoon and trying to get out. The little boy forgot his promise, and with all kindness opened the cocoon. The butterfly few out, soared up in the air, and then suddenly fell down to the ground, dead. The little boy went crying to the guru with the dead butterfly. The guru said to the little boy, 'You see, you must have opened the cocoon and helped the butterfly out because you felt sorry for it, but what you didn't realize is that the way the butterfly gets strong is by beating his wings against the side of the cocoon. And when you don't let him beat his wings that way, his wings will be weak and he must die.' "

Jacobs interprets the story by saying, "What we have to remember about the poor in America, the people with whom we are so concerned, is that our society is the cocoon against which the poor must beat to gain their strength. And the only way in which the poor will get strong is if society is willing to take the risk of being beaten against."[43]

NOTES

1. See *The Economic Absorption and Cultural Integration of Inmigrant Workers,* Lyle W. Shannon, Elaine Krass, Emily Meeks, and Patricia Morgan, University of Iowa, Iowa City, Iowa, 1964, 453 pp., mimeographed revised and completed research report to NIMH.

2. The goal of the sociologist in action programs, or planned social change as it was more frequently called in the past, has been discussed in the literature at great length. On the one hand there are those who contend that we simply do not know enough sociology, that is, enough about human behavior, to give scientific counsel to those who wish to deal with the most pressing social problems of our time. And on the other there are those who see a variety of roles for the sociologist and who have even observed the consequences of the various intervention programs designed and in some cases even carried out by sociologists. One book of readings that presents a number of articles describing sociologists in action in such areas as education, religion, health, community development, race relations, crime, political action, poverty, is *Sociology in Action: Case Studies in Social Problems and Directed Social Change,* Arthur B. Shostak (ed.), Homewood, Ill.: The Dorsey Press, 1966, 159 pp. This volume also includes an annotated bibliography of 33 books constituting a representative sample of volumes describing the intervention experiences of sociologists.

3. See James S. Coleman, et al., *Equality of Educational Opportunity,* U.S. Department of Health, Education, and Welfare, 1966, p. 41. Schools with a high concentration of Mexican-American pupils received disproportionate numbers of poorly trained teachers. An unpublished study of 1,650 elementary teachers in the lower Rio Grande Valley of Texas reported that 10 percent had no Bachelor of Arts degree, 13 percent were serving with emergency credentials, and 30 percent had provisional credentials. Only 57 percent were fully credentialed

Texas teachers (unpublished survey by A. R. Ramirez of elementary teachers in the Lower Rio Grande Valley, 1966, mimeographed). A Valley superintendent reported that he was forced to employ some teachers with as few as 60 college units of credit. In other areas the situation appears much better. The general shortage of qualified teachers throughout the Southwest thus strongly influences the quality of Mexican-American schooling.

4. The program of the Racine Environment Committee has been notable in its efforts to deal with this problem in Racine during the past three years.

5. Perhaps even more pertinent to the Racine situation is A. H. Halsey's article, "Youth and Employment in Comparative Perspective," *Poverty in America,* Margaret S. Gordon (ed.), San Francisco: Chandler Publishing Company, 1965, pp. 139-160.

6. We have no intention of splitting hairs with anyone, academic or otherwise, who suggests that we are presuming that inmigrants wish to be absorbed into the economy. That a small proportion of the population, regardless of race or ethnicity, wishes to survive without absorption (enjoy the benefits of the productivity of others) cannot be denied. We are concerned with those who have moved to Racine in order to maximize their chances of absorption.

7. For the perceptive journalist's description of how an unemployed Negro with four children faces the fact that there are no funds for a welding course that he believes will make it possible for him to secure employment, see David Pfankuchen, "Seeks New Skills To Beat Joblessness," *Racine Journal-Times,* Monday, January 18, 1971, p. 5B.

8. For a recent volume that is rather critical of the success of urban education in educating the "new immigrants," see *Urban Education: Problems and Prospects,* William M. Perel and Philip D. Vairo, New York: David McKay Company, 1969, 145 pp.

9. "The problem with vocational education has traditionally been that it has . . . served the relatively *more* prosperous . . . rather than the poorest, the *more* able rather than the least equipped, the *more* privileged rather than the deprived and disadvantaged." Francis Keppel, "Vocational Education and Poverty," *Poverty in America,* Margaret S. Gordon (ed.), San Francisco: Chandler Publishing Company, 1965, p. 136.

10. See Gloria Channon, "Bulljive—Language Teaching in a Harlem School," The Urban Review, Vol. 2, No. 4, February, 1968, pp. 5-12. School teachers who have not grown up in the inner-city should be trained to overcome their initial disadvantage in teaching children in these areas. Exactly whose responsibility it is to conduct teacher training along these lines is open to discussion, but it is possible that pre-teaching institutes could assign articles, such as Channon's, and instruct on the availability of appropriate teaching materials. Also see, by Gloria Channon, "Spelling as an Aid to Reading," The Urban Review, Vol. 3, No. 3, January, 1969, pp. 35-39.

11. See Arthur B. Shostak, "Educational Reforms and Poverty," in *New Perspectives on Poverty,* Arthur B. Shostak and William Gomberg (eds.), Englewood Cliffs: Prentice-Hall, Inc., 1965, pp. 62-69.

12. For a description of a 1965 experiment, see Robert Rosenthal and Lenore Jacobson, "Pygmalion in the Classroom," The Urban Review, Vol. 3, No.

1, September, 1968, pp. 16-20, taken from the book by the same title, *Pygmalion in the Classroom,* Holt, Rinehart & Winston. For further discussion of the subtle influence of the teacher in the classroom, see Peter and Carol Gumpert's "The Teacher as Pygmalion: Comments on the Psychology of Expectation," The Urban Review, Vol. 3, No. 1, September, 1968, pp. 21-25.

13. The teacher of students with Mexican-American antecedents should be prepared to understand their cultural background. Julian Nava in "Cultural Backgrounds and Barriers that Affect Learning by Spanish-Speaking Children," in *Mexican-Americans in the United States: A Reader,* John H. Burma (ed.), Cambridge: Schenkman Publishing Company, 1970, pp. 125-133, makes a variety of suggestions for teachers in states relatively close to Mexico who have large numbers of Mexican-Americans in their classes. The same suggestions are pertinent for any industrial community with a large proportion of Mexican-American children.

14. See Deborah I. Offenbacher, "Cultures in Conflict: Home and School as Seen Through the Eyes of Lower-Class Students," The Urban Review, Vol. 2, No. 6, May, 1968, pp. 2-8.

15. "The Invisible Minority," Report of the NEA-Tucson Survey on the Teaching of Spanish to the Spanish Speaking, Department of Rural Education, National Education Association, 1201 Sixteenth Street, N.W., Washington, D.C. 20036, 1966. The recommendations of the committee appear in abbreviated form in *Mexican-Americans in the United States: A Reader,* John H. Burma (ed.), Cambridge: Shenkman Publishing Company, 1970, pp. 103-118. The reader may also wish to refer to a series of papers prepared for the Chicano Studies Summer Institutes held in 1970 in Aztlan. These papers deal with the Chicanoism and problems and procedures of establishing a Chicago Studies program in the college or university. They were produced with the financial assistance of the National Endowment for the Humanities.

16. There is some evidence that the best bilingual teachers come from those whose native language is Spanish. This entails bringing Mexican-Americans into the school system as teachers. See Joseph Stocker, "Help for Spanish-speaking Youngsters," American Education, May 1967, pp. 17 ff. The same issue of American Education contains an excellent article by Ann Parker, "Mees, You Goin' to be a real teacher now, Don'cha?" pp. 15-16. In this article, Miss Parker, a graduate student studying English and Education at Texas College of Arts and Industries on a Woodrow Wilson Fellowship, deals with her experience at teaching the Spanish-speaking children of migrant workers in Texas. She points out that the vocabulary of the children is limited in both English and Spanish and that, unfortunately, what the children will speak eventually is "Tex-Mex," a patois formed by jumbling broken English and broken Spanish together. As she puts it, this will enable them to live in the world of the migrant worker and no other. Instead of being bilingual they are non-lingual, for their language is not a recognized one.

17. See Chris Tree, "Schools in Storefronts," The Urban Review, Vol. 2, No. 4, February, 1968, pp. 16-18.

18. For a series of papers discussing the problems of education in lower socioeconomic status areas, see *Education in Depressed Areas,* A. Harry Passow (ed.), New York: Teachers College Press, 1963, 359 pp. Particularly see Henry

Saltzman, "The Community School in the Urban Setting," pp. 322-331. Saltzman suggests that a school should be built whose facilities would be available for use by social agencies, health agencies, and interested citizens groups on a seven-days-a-week, 12-months-a-year basis. This sort of school, as Saltzman sees it, would be the center of the community.

19. See Frank A. Petroni, " 'Uncle Toms': White Stereotypes in the Black Movement," Human Organization, Vol. 20, No. 4, 1970, pp. 260-266.

20. There has been considerable discussion of the advisability of lowering admissions requirements for so-called disadvantaged students with the idea that once in the school they will in one way or another overcome their earlier disadvantage. The contention is that if there is strong motivation to achieve academically it will be, although perhaps dormant for many years, awakened in college. Julian C. Stanley in "Predicting College Success of the Educationally Disadvantaged," Science, Vol. 171, No. 3972, 19 February, 1971, pp. 640-647, contends that admission to college should be based substantially on test scores and high school grades. His definition of the disadvantaged is not based on race or ethnicity. As he puts it, "From the standpoint of an admissions officer, the educationally disadvantaged applicants to his college could be simply those who, on the basis of all available information, including high school grades, test scores, socioeconomic status, race, ethnic origin, and available financial support, are likely to have appreciably more academic difficulty than the typical minimally admissable student. Thus, the son of a distinguished alumnus is educationally disadvantaged for his father's college if it is predicted that he will fail most of his courses and not persist to graduation. The valedictorial of a large high school, who has a sizeable national scholarship but whose parents are illiterate and penniless, cannot, by this criterion, be considered greatly disadvantaged educationally." He goes on to point out that this definition is certainly not in accord with current use in the literature. For a critique of the open admissions policy at the City University of New York, see "Open Admissions: A Critique," Community, Vol. II, No. III, April, 1970, pp. 6-7.

21. Throughout the Southwest there is an 80 percent drop-out rate for Mexican-Americans who do enter college. In San Diego State College's High School Equivalent Program (HEP), the goal is to develop capacity to pass the required college entrance exam. See Urban Affairs Newsletter, Vol. III, No. 1, February, 1971, p. 2.

22. The SEEK Program (Search for Elevation, Education and Knowledge) officially began in 1966 with approximately $1 million from the New York State Legislature. Its aim was to assist minority group students from poverty areas in New York City. The ultimate goal would be the admission to the City University. SEEK was conceived as a panacea to produce students capable of handling college-level work. By 1969 there were more than 4,000 students in the program. The program included tutorial aid for students who showed deficiencies in mathematics, English, science, or any other course in which a student might have had difficulty. See The SEEK Program: A SEEK Student's View, by Jackie Robinson, Community Issues, Institute for Community Studies, Queens College, Vol. 2, No. 3, July, 1970, 16 pp., which enables one to see how students perceive the organization of the program and how it contributed to the difficulties encountered. The view of Black students, whether we agree with

them or not, must be taken into consideration if a program is to have any chance of success. One quotation from the Robinson brochure gives some indication of the nature of the problem: "The whole concept of education must be changed for Black people. And the only way this can be done is to create Black educational complexes which address themselves to the needs of Black people. Educational institutions (elementary, secondary and higher education) must be erected, evaluated, and perfected by Black people. Programs such as SEEK are only weak attempts at recitifying the educational wrongs committed against Black people.

"Black educational systems must be instituted to provide identity, purpose, and direction to Black people. White institutions have only succeeded in distorting our values and superimposing alien and white values upon Black minds. For the most part, education for Black people has only prepared them to accept white collar positions that inevitably make them lackeys to a Judeo-Christian society. Negro institutions such as Howard and Fisk are no exceptions, for they are mere facsimiles of white colleges."

23. See *Education for the Urban Disadvantaged: From Preschool to Employment,* A Statement on National Policy by the Research and Policy Committee of the Committee for Economic Development, March, 1971, particularly the "seven imperatives" for education—for example, preschooling is desirable for all children but a necessity for the disadvantaged.

One new program that should be given serious consideration is the Home Start Program now being sponsored by the Department of Health, Education, and Welfare in Waterloo, Iowa. As its name implies, children in Home Start receive their first educational experiences in the home. The program lasts for three years and terminates when the children reach kindergarten. Preliminary evaluation shows that the children have gained about 10 I.Q. points. Des Moines Sunday Register, Picture Magazine, March 14, 1971, pp. 4-7. Also see Chapter 5, "Schools: Broken Ladder to Success," *Spanish Harlem: Anatomy of Poverty,* Patricia Cayo Sexton, New York: Harper & Row, 1965, pp. 47-70.

24. Although it should almost go without saying, one cannot expect to even know where to start a program of planned social change or intervention or assistance to the less fortunate in the community unless one understands the relative importance of various social institutions. For example, the position of the church, as we have said before, changes over a period of time and what has been said about the Negro church in the South is not necessarily applicable in Racine. However, to the extent that the Negro population consists of persons who are relatively recent inmigrants from the South, then an understanding of the variety of Negro congregations that may be present in the community is important. For a good short chapter on social institutions in Negro communities, see Chapter 5 of Alphonso Pinkney's *Black Americans,* Englewood Cliffs: Prentice-Hall, Inc., 1969.

25. See Harvey M. Choldin and Grafton D. Trout, *Mexican-Americans in Transition: Migration and Employment in Michigan Cities,* Department of Sociology, Rural Manpower Center, Agricultural Experiment Station, Michigan State University, 1969. In their comments on guided migration, Chapter 6, "Policy and Research Recommendations," they suggest, among other things, that given adequate employment opportunities, migration of Mexican-Americans

should be encouraged and facilitated to smaller or middle-sized communities rather than major metropolitan areas. They also suggest that preparation for migration should be accomplished in the place of origin where feasible as well as in place of destination. The introduction of ex-migratory, experienced, educated bilingual personnel into programs of assistance to migrants is also recommended. The last of seven recommendations on the resettling of migrants is that some form of income maintenance be instituted during the first year or so of settlement. The authors also make a number of recommendations regarding those who have been residents for several years, including the use of bilingual instructors in job training programs and greater emphasis on in-plant training by employers as contrasted to general adult education. Also see Niles M. Hansen and William C. Gruben, "The Influence of Relative Wages and Assisted Migration on Locational Preferences: Mexican Americans in South Texas," Social Science Quarterly, June, 1971, Vol. 52, No. 1, pp. 103-114. The authors conclude, on a basis of 1,043 interviews with South Texas junior and senior high school students, that comprehensive relocation assistance would be more efficient in increasing income and employment opportunities than efforts to attract industry to the area.

26. For an excellent criticism of the United States Employment Service, see Edward T. Chase, "The Job-Finding Machine: How to Crank It Up," *New Perspectives on Poverty,* Arthur B. Shostak and William Gomberg (eds.), Englewood Cliffs: Prentice-Hall, Inc., 1965, pp. 83-94. As Chase points out, "Indeed USES does not even know, on a national basis, where the job openings exist or are likely to occur, or the location, and the skills or lack of them, of the unemployed men and women who might fill those vacancies."

27. S. M. Miller and Frank Riessman, "The Working Class Subculture: A New View," Social Problems, Summer, 1961, p. 95.

28. The Institute for Research on Poverty at the University of Wisconsin has reprinted a multitude of excellent papers on poverty, some of which evaluate programs that are supposedly operating for the disadvantaged.

29. Perry Levinson, *Chronic Dependency: A Conceptual Analysis,* Research Working Paper No. 1, U.S. Department of Health, Education, and Welfare, 1964, p. 6. Although Levinson's analysis provides extensive information for use by social work administrators, we have only dealt with one section of his detailed analysis.

30. See Robert C. Hanson and Ozzie G. Simmons, "Differential Experience Paths of Rural Migrants to the City," Chapter 6 in *Behavior in New Environments: Adaptation of Migrant Populations,* Eugene B. Brody (ed.), Beverly Hills: Sage Publications, 1970, pp. 145-166.

31. Students in the St. Louis University School of Medicine, through its Department of Community Medicine, learn to view the patient in his total context in eight different "ecologies" ranging from white suburban to all-black central city. See Marilyn E. Ludwig, "Medical Care in a Community Context," City, March/April, 1971, pp. 30-32.

32. Walter B. Miller, "Implications of Lower-Class Culture for Social Work," The Social Science Review, Vol. XXXIII, September, 1949, pp. 20-21.

33. Walter Miller, op. cit., pp. 28-29.

34. A number of specific suggestions by Daniel Rosenblatt in "Barriers to

Medical Care for the Urban Poor," *New Perspectives on Poverty*, Arthur B. Shostak and William Gomberg (eds.), Englewood Cliffs: Prentice-Hall, Inc., 1965, pp. 69-76, indicate that health, education, and welfare services should be integrated on the local level. The multiplicity of health and welfare services in Racine County suggests this as a step for consideration. He also suggests the introduction of the health visitor, who is an indigenous person serving as translator, advance contact, and a cultural bridge between modern medicine and traditional notions of how illness should be dealt with. Another suggestion worth considering is the concept of satellite clinics.

35. See Vera M. Green: "The Confrontation of Diversity within the Black Community," Human Organization, Vol. 29, No. 4, Winter, 1970, pp. 267-272. It is just this historical and contemporary diversity as reviewed by Green that makes it such folly to address programs to the Negro community as though those families consist of unmarried women with children. Green points out that the unmarried Negro family has so frequently been considered the norm that some Negro students feel guilty because they have fathers at home supporting the family. She goes on to point out that "needy stable Black families who require assistance are frequently overlooked in favor of multiproblem families with less urgent requirements, but who fit the stereotype of needy Blacks." For a very recent analysis of the status of the Negro family based on U.S. Census data, see Reynolds Farley and Albert I. Hermalin, "Family Stability: A Comparison of Trends Between Blacks and Whites," American Sociological Review, February, 1971, Vol. 36, No. 1, pp. 1-17. Most nonwhite families are husband-wife families (74 percent in 1969) and the majority of the nonwhite children live with both parents.

36. The reader may find *Improving the Public Welfare System*, Committee for Economic Development, New York: 477 Madison Avenue, April, 1970, quite relevant. It outlines changes in our national welfare system that are designed to not merely increase the humanitarian function of the system but to direct those on welfare in such a way that they will eventually be absorbed into the economy. For an excellent analysis and commentary on our state and federal welfare system, see Sylvia A. Law, "Crisis in Welfare," Civil Liberties, July, 1971, No. 279, pp. 1, 6 and 7.

37. We refer here to "class" subculture as well as "ethnic" subcultures.

38. Florence R. Kluckhohn, "Cultural Factors in Social Work Practice and Education," Social Service Review, 25 (March, 1961), p. 42.

39. For an excellent chapter on voluntary associations among Spanish-Americans of New Mexico, see Chapter 5, "Voluntary Associations," Nancie Gonzalez, *The Spanish-Americans of New Mexico: A Heritage of Pride*, Albuquerque: University of New Mexico Press, 1969, pp. 86-115. Also see Chapter 7, "The Effects of Urbanization," and Chapter 8, "The Continuing Scene: Activism in New Mexico: 1966-1969."

40. See Ornati's essay, "The Spatial Distribution of Urban Poverty," in *Power, Poverty, and Urban Policy*, Warner Bloomberg, Jr. and Henry J. Schmandt (eds.), Volume II of Urban Affairs Annual Reviews, Beverly Hills: Sage Publications, 1968, 604 pp.

41. A short summary of changing strategies of Negroes in the United States in an effort to reach their goals may be found in Chapter 4, "Action and

Integration," of *Transformation of the Negro American,* by Leonard Broom and Norval Glenn, New York: Harper & Row, 1965, pp. 55-80.

42. For an early evaluation of Community Action Programs, see *A Relevant War Against Poverty: A Study of Community Action Programs and Observable Change,* New York: Metropolitan Applied Research Center, Inc., 1968, 275 pp. The Community Action Programs of 12 cities were visited and studied in depth. This volume gives some idea of the variety of programs that have been planned and the mixed responses given to them by people in the community. Programs in New Haven, Syracuse, New York, Patterson, and Minneapolis were considered relatively effective, while those in Chicago, Cleveland, and Boston were rated as relatively ineffective. The importance of agreeing upon how to evaluate effectiveness of a program at the outset and evaluation by outsiders periodically cannot be underestimated.

43. See Paul Jacobs, "Discussion," *Poverty in America,* Margaret S. Gordon (ed.), San Francisco: Chandler Publishing Company, 1965, pp. 413-417.

TEN YEARS LATER

Preparations for a Re-study of the Racine Respondents

The possibility of a re-study of the Racine respondents, wherever they might be at some later date, was always in mind. If each of the 973 respondents was to be re-interviewed we would have 282 Mexican-Americans, 280 Negroes, and 411 Anglos. These samples would be larger than in any previous year and would permit more partialling. The problems that might arise in combining the 1959 and 1960 samples could be surmounted and questions of randomness would be overshadowed by the advantage of the increased number of respondents in the most relevant age and time in the community groups. If either the original respondent or his or her spouse could be located, then in our estimation a re-study would be quite feasible and contribute far more to the understanding of the processes of economic absorption and cultural integration than the research of 1959, 1960, and 1961.

With some forethought we had decided to apply for a small grant from the United States Public Health Service in order to pursue more intensive analyses of the data in hand. The further we progressed, the more we realized the need for sufficient funds for a thorough re-study. Knowing that an application for a larger grant would entail a longer time for processing and paying, we applied to the Public Health Service for both a large and a small grant simultaneously in July of 1969. Meanwhile, we began to search for the 973 respondents and to gather information about their status circa 1970 for the concluding chapter of the present volume. Perhaps any early success on our part might favorably affect the deliberations of the agency from which we were seeking funds. Judy McKim (Project Coordinator for the Iowa Urban Community Research Center) was given responsibility for locating the respondents and assisting the Director of the Center in preparing for a re-study.

The first move in the process of locating our former respondents was the formulation of a short Spanish-English questionnaire and a letter (Spanish and English for Mexican-Americans and English for Negroes and Anglos) varying in form if the respondent had been interviewed in 1959, in 1959 and 1960, in 1960, or in 1960 and 1961.[1] Both were mailed to respondents between the first week of November and the second week of December of 1969, using the address most recently given by the Racine City Directory or the Racine Telephone Director or, lacking that, the address given at the time of the last interview.

Needless to say, the results were disappointing. During the ensuing month only 113 completed questionnaires were received while returns of undeliverable letters were numerous. Our dismay was tempered somewhat in that many of the questionnaires were accompanied by lengthy letters in which respondents described either their unfortunate experiences in the community or, less frequently, their successes and upward mobility.

In December, the Applied Research Branch of the Center for Studies of Mental Health and Social Problems of the National Institute of Mental Health indicated their interest by raising a variety of questions about the re-study, among them the difficulty or perhaps impossibility of locating the 973 respondents. At that point we had located nine percent of the respondents and we knew that 17 percent were not at the address to which we had sent a letter, but we were not sure whether the remaining 74 percent had been delivered in Racine or forwarded.

It was evident that a more diligent search was in order. Continually referring to city directories and telephone directories, we repeatedly checked and re-checked all names mentioned in the original interviews in an effort to find clues as to the correct address of respondents. This process involved some imagination; not only had some names (given and surname) been misspelled on the interview schedules, on lists of completed interviews, and in the directories, but original names had been changed by remarriages or re-spelled by choice. In some cases we went so far as to send a letter to the hometown of a respondent with a request that it be forwarded.

In February of 1970, further discussion was held with the Applied Research Branch in reference to our progress. Although relatively few of the questionnaires had been returned, we indicated that the picture was more reassuring since the addresses of 66 percent of the Mexican-American respondents, 79 percent of the Negroes, and 79 percent of the Anglos had been found in 1969 Racine directories. Inadvertently during this communication we learned that our small grant application had been approved but not paid. It was determined that funds from the small grant award could be utilized not only for further analyses of the original data but also in further search for our respondents. In May we received the grant, totaling $7,200, for the period June 1, 1970 through May 31, 1971—a welcome impetus to the project.[2] Twelve University of Wisconsin-Parkside students were trained to telephone and call upon respondents in order to verify addresses and encourage the return of the questionnaire. If a respondent requested assistance in completing the questionnaire, the students were instructed to give whatever aid was needed.

During the summer of 1970, six trips to Racine were necessary to closely supervise the work of the students. Concurrently, new Racine directories were consulted, as were telephone and city directories for Kenosha and Milwaukee. City, county, state and federal agencies assisted us in every manner possible in finding addresses of respondents. The Racine Unified School District school census was particularly useful at this stage of the research, since identification was positive whenever the names and ages of children in the Census could be matched with those given in previous interviews. Inquiries were made of neighbors and relatives—a task which proved arduous. Many of our respondents had names that were similar to or identical to those of other persons in the community; others did not wish to be found. Houses had been eliminated as new thoroughfares were constructed. Female respondents had remarried. Some had simply changed their names for the sake of convenience; others were institutionalized or deceased. Letters were written to the children of respondents, to their families, and to their friends. We telephoned former places of residence. Slowly addresses were verified and questionnaires returned. Each month we sent a progress report to interested persons in the National Institute of Mental Health. By September, 1970, we had actually located 64 percent of the Mexican-Americans, 71 percent of the Negroes, and 81 percent of the Anglos. By November, we succeeded in reaching 78 percent of the Mexican-Americans, 88 percent of the Negroes, and 88 percent of the Anglos.

In December we corresponded with acquaintances in the Texas communities where our Mexican-American respondents had formerly resided, seeking knowledge of the persons on enclosed lists and indicating that we would contact them personally between Christmas and the New Year. We visited every community along the border between El Paso and Brownsville, as well as San Antonio, Crystal City, Cotulla, and Pearsall. We were given unqualified assistance in finding our respondents; the mayor of Cotulla alone devoted half a day to the search. By January we were able to report that 85 percent of the Mexican-Americans and 93 percent of both Negroes and Anglos had been located. Shortly after our return we learned that the larger application for a three-year re-study of the absorption of migrant workers had been approved and funded.[3]

The Mobility of Respondents 1960-1970

The cut-off point for analysis of the returned questionnaires and the spatial mobility of respondents was April 7, 1971. The status of the 973 respondents as of that time is shown in Table 63. Some problems are impossible to avert, such as: the possibility that a few of the persons may not be the ones interviewed originally but possess names identical to them; the problem of deceased respondents; and what may be called double-deceased families. We shall, of course, continue the search and in some cases we will be forced to interview the children in order to see how their parents played out their lives. Why so few Negroes returned their questionnaires cannot be readily explained; perhaps they were reluctant to disclose their status because unemployment was very high among Negroes in Racine at that time.[4] But unemployment was also high among the Mexican-Americans.

TABLE 63. STATUS OF COMBINED 1959, 1960, AND 1961 SAMPLES: APRIL 1971

	Mexican-American	Negro	Anglo	Total
Mailed questionnaire returned	116	29	205	350
Address verified, no questionnaire returned	131	234	171	536
Deceased	7	2	19	28
Not located	26	15	18	59
Total	280	280	413	973

Our first tentative evaluation of the Mexican-American, Negro, and Anglo samples is presented within an ecological perspective. How has the spatial distribution of each of these groups changed in a sociologically meaningful fashion between 1960 and 1970?

Let us first turn to Table 64, which reveals the balance of movement by race and ethnicity from 1960 to 1970. The reader will remember that we selected our 1960 samples on a basis of the race and ethnic composition of each public school attendance center in Racine. Thus we had a target population for each area which was based on the proportion of Mexican-Americans, Negroes, and Anglos in each.[5] The sampling fraction differed for each race or ethnic group—one Mexican-American out of every six was to be inter- viewed, one Negro out of every 10, and one Anglo out of every 100. In our present discussion when we refer to movement between 1960 and 1970 we mean that 282 Mexican-American families followed such and such pattern of movement, likewise 280 Negro families and 411 Anglo families (except for those whom we have not located).

It might be wise for the reader to first examine Map 3, which shows how attendance centers have been modified and divided to increase the homogeneity of areas. Instead of the 19 areas utilized in our earlier years of research for sampling purposes as shown in the map in Chapter 1, we now have 24. The boundary lines were modified on a basis of careful study of the 1960 Census Block Data, examination of the ecology of Racine based on Guttman scale scores for each block, geometric scale scores for each block, consultation with administrators in the Unified School District, and direct observation of the physical environment during hundreds of miles of driving in Racine.

Group V consists of the seven modified attendance centers that are predominantly Anglo and which have experienced the least population change since 1960: Jerstad-Agerholm, Roosevelt, East Winslow, East and West Fratt, West Jefferson-McKinley, and South S.C. Johnson. These areas were practically devoid of Mexican-Ameri- cans and Negroes in 1960, but by 1970 a few dozen families were living there. Among them were eight of our Mexican-American and six of our Negro respondents. All save West Jefferson-McKinley were also in the top or next to the top categories on a five-point characteristics of housing scale based on 1960 U.S. Census data. All except East Winslow are separated from the inner city by natural geographic barriers or intervening school attendance centers.

Three of the six modified attendance centers in Group IV

TABLE 64. DISTRIBUTION OF COMBINED 1959, 1960 AND 1961 SAMPLES IN CIRCA 1960 AND 1970[1]

Districts and Groups of Districts[2]	Mexican-American 1960	Mexican-American 1970	Negro 1960	Negro 1970	Anglo 1960	Anglo 1970
Group V						
Jerstad-Agerholm	1	0	3	4	25	23
Roosevelt	0	0	0	0	24	19
E. Winslow	2	1	0	1	7	11
W. Fratt	0	0	0	1	25	24
W. Jefferson-McKinley	2	2	0	0	10	12
S. S.C. Johnson	1	3	0	0	4	10
E. Fratt	0	2	0	0	11	12
	6	8	3	6	106	111
Group IV						
Wadewitz	2	5	0	1	12	9
W. Knapp	2	4	0	0	12	21
N. Lincoln	1	4	1	3	3	3
E. Knapp	0	3	0	1	11	10
E. Mitchell	2	5	0	0	7	10
W. Mitchell	0	4	0	1	23	17
	7	25	1	6	68	70
Group III						
Janes	13	8	4	3	11	12
Washington	14	17	2	12	17	12
N. S.C. Johnson	11	13	0	10	29	15
Lakeside	13	7	9	10	0	0
	51	45	15	35	57	39
Group II						
W. Winslow	14	8	10	9	14	10
Howell	20	8	70	44	12	8
Hansche	58	41	22	4	1	0
	92	57	102	57	27	18
Group I						
W. Stephen Bull	7	11	44	47	30	5
E. Jefferson	13	15	15	24	10	3
Franklin	26	13	32	26	47	21
Garfield-S.Lincoln	74	41	68	52	65	27
	120	80	159	149	152	56
Total in Five Areas	276	215	280	253	410	294
Outside Racine	4	32	0	10	3	82
Not Located or Deceased	0	33	0	17	0	37
	280	280	280	280	413	413

1. Residential location of all for 1960 except those interviewed only in 1959.
2. Districts and areas as shown on Map 3, Appendix B, highest socio-economic status to lowest.

[325]

(Wadewitz, West Knapp, and East Mitchell) were in the fourth characteristics of housing category, while one (West Mitchell) was in the highest; North Lincoln plus East Knapp were in the middle category. The proportion of Mexican-American respondents in these areas has increased over threefold in the 10-year period. That of the Negroes has increased somewhat but the Anglos remained essentially the same except in West Knapp, where the proportion increased.

The third area (Group III) consists of four modified attendance centers: Janes, Washington, and Lakeside were in the next to the lowest characteristics of housing category and North S. C. Johnson was in the middle. These districts contained 18 percent of the Mexican-Americans in our sample in 1960 and slightly less in 1970. But while only five percent of the Negroes were in this area in 1960, there were over 12 percent in 1970, with considerable movement into Washington and North S.C. Johnson. Only half the Anglo respondents were left in North S.C. Johnson in 1970.

Group II comprises West Winslow, Hansche, and Howell, the first two being in the lowest characteristics of housing category, while Howell was in the next to the lowest. In 1960, Howell was already heavily Negro while Hansche and West Winslow were Negro and Mexican-American. The number of Mexican-American respondents in this area decreased by one-third and the Negro by almost one-half between 1960 and 1970.

The last group of modified attendance centers (West Stephen Bull, East Jefferson, Franklin, and Garfield-South Lincoln) contained more Mexican-Americans and Negroes in 1960 than any other; by 1970, it had lost all of our Mexican-American and two-thirds of our Anglo respondents, but the number of Negro respondents had only slightly declined. Three of the four centers in this area were in the lowest characteristics of housing category.

In 1960, only three Anglos in the original sample lived outside the Racine areas to which we have referred; by 1970, 82 had moved to the suburban areas or to other communities. Just 32 of the Mexican-American sample moved to communities outside of Racine and, insofar as we can determine, only 11 of these have returned to Texas. Most startling of all—just 10 Negroes have left Racine. These figures may change somewhat as the remaining 59 persons are located.[6]

The residential mobility that took place among our Racine respondents between 1960 and 1970 is shown in Table 65 in simplified form. Of the 103 Mexican-Americans who resided in the

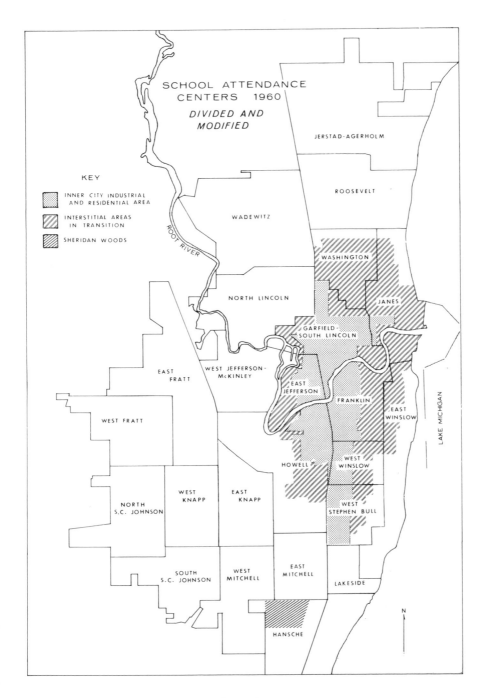

FIGURE 7.

TABLE 65. RESIDENTIAL LOCATION OF MEXICAN-AMERICANS IN 1960 AND 1970[1]

1960-1970	V	IV	III	II	I	Outside Racine	Other Midwest	Texas	Total
V	3		1		1	1			6
IV	1	3	1		1			1	7
III	3	5	17	4	8	3	2	4	46
II	1	2	9	44	17	4	2	4	83
I		15	17	8	52	4	4	2	102
Outside Racine				1	1	1			3
	8	25	45	57	80	13	8	11	247

1. Total for areas as presented in Table 64.

heavily minority group populated areas (Group I) in 1959-1960, 52 remained and 50 moved to better residential areas by 1970 (movement to better residential areas is movement to the left on the table) but *none* as far up the scale as the suburban areas in the top characteristics of housing category. Of the 84 who lived in Group II, 44 remained, 17 moved down into Group I, and 12 moved upward, i.e., into Groups III, IV, and V. If the movement to suburban areas outside Racine is included, then the upward equals the downward movement. When the middle interstitial group of centers is observed, of 46 who resided there in 1960, 17 remained and equal numbers moved upward to the suburbs or downward. Overall, the movement that took place among Mexican-Americans in the 10-year period showed that only 37 of the 233 Mexican-Americans moved out of the lowest housing categories and into the two highest categories or to the suburbs outside of Racine.

Residential moves by the Negro sample as shown in Table 66 were more limited in scope. Thirty-six of the 148 Negroes who formerly resided in Group I moved upward, 31 of them into the two next highest groups where there were already sizeable Negro populations in 1960. When the second group of modified centers is considered (Group II), the movement from it is downward into the lowest area for 36 of 97 families while only 18 moved upward or into the suburbs. The propensity of our Negro sample to remain in the inner city is verified by the fact that only seven of 245 who resided there in 1960 moved into the top two groups (Groups IV and V) of Racine by 1970.

The Anglo pattern of movement (Table 67) contrasts sharply with those just described for Mexican-Americans and Negroes. Of those

TABLE 66. RESIDENTIAL LOCATION OF NEGROES IN 1960 AND 1970[1]

1960-1970	V	IV	III	II	I	Outside Racine	Other Midwest	Total
V	2				1			3
IV					1			1
III	1	2	5	2	4			14
II		2	13	41	36	3	2	97
I	3	2	17	14	107	3	2	148
	6	6	35	57	149	6	4	263

1. Total for areas as presented in Table 64.

128 who lived in the lowest area in 1960, only 50 remained; 33 of the 45 who moved up had gone into the two highest areas. In the other four groups, the pattern of movement of those who left was also upward.

We might also perceive the movement of our samples in terms of the ethnicity of the areas they moved from and those they entered. Mexican-Americans left areas that contained Mexican-Americans and Negroes and moved into other areas already containing persons from both groups, with the exception of those who moved into Group IV. Negroes moved from heavily populated Negro areas to other areas already containing sizeable numbers of Negroes, with the exception of the Washington and North S.C. Johnson areas. But Anglos, by contrast, moved out of areas that contained sizeable Negro and Mexican-American populations and into areas that were predominantly Anglo.

Our second type of appraisal involves an analysis of change that has taken place in the lives of the persons in the samples, based on

TABLE 67. RESIDENTIAL LOCATION OF ANGLOS IN 1960 AND 1970[1]

1960-1970	V	IV	III	II	I	Outside Racine	Other Midwest	Other U.S.	Total
V	76	5	1	1	3	8	4	3	101
IV	7	42		1	1	10	2	3	66
III	4	4	32		1	7	2	2	52
II	7	3		10	1	2	2		25
I	17	16	6	6	50	20	10	4	129
Outside Racine						2	1		3
	111	70	39	18	56	49	21	12	376

1. Total for areas as presented in Table 64.

those who returned the mail questionnaire. Since only 40 percent of the mail questionnaires were returned, with the Negro percentage very small in comparison to those for Anglos and Mexican-Americans, a detailed analysis of the results will not be presented. However, we must note that while Anglo movement was upward in terms of occupational levels, only half of them *believed* that the job they had in 1970 was better than the one held previously. Few Mexican-Americans had actually moved upward though one-third *thought* that they had better jobs in 1970 than in 1960. Among the few Negroes on whom we had data there appeared to be some mobility and almost half thought that they had better jobs than in 1960. Anglo upward movement was more definitive than that of the other groups, with twice as many or more in every occupational category moving upward rather than downward. Not so for the Mexican-Americans. Some moved from operative to industrial labor while others went from industrial labor to operative. For the few Negro returns, movement was upward, from operative to craftsman or foreman from industrial labor to operative, but again their mobility could not be compared with that experienced by Anglos.

Reference to national population statistics at this point served to further emphasize the position of the Mexican-American in the United States. The occupational distribution of Mexican-Americans in the United States in 1960 was the lowest of all ethnic groups. Only 7.9 percent reported positions at the professional, technical level in contrast to 14 percent for the entire U.S. Seven and four-tenths percent were managers, officials, and proprietors, while 14.2 percent was reported for the nation; 16.6 percent of the Mexican-Americans were laborers, agricultural and industrial, while 8.8 percent of the United States were in this category.[7] These figures ought not to be too surprising since only 80.2 percent of the Mexican-Americans reported that they could read and write English in comparison with 95 percent for all persons in the United States.[8] Their median years of school completed was the lowest. Only 5.3 percent had graduated from college in comparison with 15.2 for the United States among persons 25 years of age or older.[9]

Changes in the United States and in
Racine Between 1960 and 1970

These statistics take on additional meaning when we refer to income data. Among those respondents in the sample who returned their mailed questionnaire, Anglo income increased from the modal category of $6,000-$6,999 to $10,000-$10,999, while Mexican-American income went from $5,000-$5,999 to $8,000-$8,999. The 18 Negroes who returned the questionnaire increased their median income of $6,000 to $7,500.

U.S. Census data showed white median family income in the North and West increasing from $6,247 in 1959 to $10,202 by 1969, a change which is comparable to our Anglo sample return given above.[10] In a more recent survey, the median income for 1970 for all white families in the north-central United States was set at $10,508.[11] In 1959, Negroes in the North and West had a median family income of $4,407, which increased to $7,408 by 1969;[12] The median income of Negro families in the north-central states by 1970 was $7,718.[13] These figures are not consistent with the Racine sample—but the difference may be because the Negroes who responded to our questionnaire were not representative of Negroes in Racine. The median income of heads of the household with "Spanish" origin was $5,641 for the United States in 1969, a figure not comparable to our sample, since it represents all persons, is disproportionately "Spanish," and disproportionately rural.[14]

Conclusion

The reader may presume that by this time the inmigrants of the 1950's as well as their children have been absorbed into the economy of Racine and become so well integrated into the larger society that such problems are no longer a concern of that community. While those who came at an earlier period have experienced limited absorption and integration, others meanwhile are newly arriving so that the problem never ceases. The continuous nature of the process was brought home to us as we interviewed a number of the more recent inmigrants during the spring of 1971 while pretesting the new interview schedule.

Most descriptive of the adjustment problems of a more recent inmigrant was the interview with Olivia. Our interviewer will never forget the experience. She told us many times that although her respondent lived only three blocks from her home, she had not realized that people lived under such depressing conditions in Racine nor that there were people in such distress in her own neighborhood. Just getting to the apartment where Olivia and Oscar lived made her apprehensive because the flight of steps and upstairs porch were rickety and dangerously close to collapse.

Olivia was an excellent respondent and seemed quite pleased just to have someone willing to listen to her family's problems, although she was often in tears as she talked.

The apartment was small and very drafty on even the warm spring day; it apparently lacked an independent source of heat. They lived in four rooms and shared the bath with other families in the building. As for furniture, there was none but the barest of essentials—all held together with black tape. The family had none of the items we have used in measuring level of living.

Oscar was a sharecropper in Mississippi before moving to Racine in 1967 and beginning work in one of its major industries, a job obtained through his cousin. Although the work had been steady at first, he was soon laid off and could find no other job because he could neither read nor write. The family had been on welfare eight months. General assistance from the welfare department was supplemented by his wife's income from part-time housework. Olivia spoke with great sadness as she described the reaction of others in the community (and particularly in her church) to the family's position. She was bitter about the difficulties that her daughter (one of three children) was having in school with her classmates.

She told the interviewer that if only they could get the money together the family would return to the South—where being on welfare was no shame because just about everyone in the community was on it. When asked a series of questions about where Negroes were best off in the United States, all of her responses were: "in the South."

At the same time, in spite of her discouragement, her sorrow, her inability to scarcely speak without giving way to tears, Olivia was able to say that she would be disappointed if her children did not obtain a college education and that she was sure that God would make it possible for her children to have one.

Olivia answered the world view questions in what would be judged a passive way but remarked "if you don't have high hopes you cannot make it in this world." Again she reminded us that welfare was a "dirty word" and just wished that they could get off it.

On questions about equality of treatment in restaurants and bars in Racine, Olivia could offer no opinion since she had not been in a restaurant in Racine nor had she or her husband been in a tavern. Neither drank, as a matter of fact.

This brief reference to a newly arrived inmigrant worker highlights the tremendous difference between the way of life of those still in the migrant cycle and those who have successfully settled in urban industrial communities, however difficult this transition may have been. It is our hope that an understanding of the processes of economic absorption and cultural integration will assist those who are concerned about the adjustment problems of inmigrants. Whatever a study of the Racine sample of 10 years ago may have contributed, it will be greatly augmented by what we shall know at the conclusion of the re-study that is now under way.

NOTES

1. The questionnaire and a sample of the letters are shown in Appendix F. It should be noted that we had planned to conduct the re-interviewing during the summer of 1970, but due to lack of funds were unable to do so.

2. The Absorption of Migrants, MH 18196-01, Center for Studies of Metropolitan Problems, National Institutes of Mental Health.

3. A Re-Study of the Absorption of Migrant Workers, MH 13321-01, Applied Research Branch, National Institute of Mental Health.

4. The number of persons below the poverty line increased by 1.2 million between 1969 and 1970. Among the Negroes, this represented an increase from 31.2 percent of the Negro families in the U.S. to 32.4 percent. In the central cities 32.7 percent of the Negroes were below the poverty line compared to 7.8 percent for whites. Current Population Reports: Consumer Income, Poverty Increases by 1.2 million in 1970, Series P-60, No. 77, May 7, 1971, pp. 4 and 6.

5. The Racine Unified School District was kind enough to provide us with race and ethnic runs of the population age 21 or over by quarter sections for 1970. Had this information been available to us in 1960, we would have stratified by quarter sections rather than attendance centers.

6. Between 1960 and 1970 the total number of "white" persons in Racine had increased from 84,332 to 84,667 according to the United States Census. Against this increase of less than half of one percent should be placed an increase of 4,738 to 10,008 for Negroes, or over 111 percent.

7. Current Population Reports: Population Characteristics. Characteristics of the Population by Ethnic Origin, November, 1969, Series P-20, No. 221, April 30, 1971, p. 23.

8. Ibid., p. 18.

9. Ibid., p. 19.

10. Current Population Reports: Special Studies, Differences Between Incomes of White and Negro Families by Region, 1969 and 1959, Series P-23, No. 35, March 10, 1971, p. 4.

11. Current Population Reports: Consumer Income, Median Family Income Up in 1970, Series P-60, No. 78, May 20, 1971, p. 4.

12. Current Population Reports: Special Studies, op. cit., p. 4.

13. Current Population Reports, Consumer Income, Median Family Income Up in 1970, op. cit., p. 4.

14. Current Population Reports: Population Characteristics, op. cit. p. 22.

APPENDICES

SAMPLING PROCEDURE

1959

A total of 386 names was listed and placed on IBM cards. An initial sample of 50 percent of those eligible was selected. A second sample of 25 percent was selected in the same way. The original listing of 386 was reduced somewhat by persons who were never married, heads of households. In the end, the 75 percent sample consisted of 256 male or female heads of families or female spouses eligible for interviewing, i.e., Mexican-American, ever married (broadly defined), heads of households or spouses of heads of households.

The Anglo sample consisted of every n'th dwelling unit within the boundaries of five areas of Mexican-American concentration. The relative equality of each area was checked in terms of the number of blocks and number of street numbers. Cards were punched from the 1958 Racine City Directory containing the addresses of dwelling units within these areas *not* having Spanish surnames. A five percent random sample was drawn from each area.

Of 335 addresses selected, 286 proved to be eligible dwelling units. Commercial units, Mexican-American and Negro households, and dwellings of single persons, never married, were eliminated. Respondents were designated as to sex by the field office.

1960

The 1960 sample was a stratified, systematic sample. It was possible to secure a list of all children, ages 0-21, with race or ethnic identification. There were 32,763 names on the list. Names on the list were separated according to the subcategories, Negro, Mexican-American, and the other cards or "Anglos"; names were also classified according to school district.

The sampling fraction by school district was one in 10 for Negroes, one in six for Mexican-Americans, and for other Racine residents, one in 200. For all practical purposes this meant that more than one in 10, one in six, or one in 200 households would be interviewed, since each Mexican-American, Negro, and other family had more than one child. The outcome of the sampling procedure was that three samples were drawn, each a systematic sample of the particular ethnic or racial group, stratified by school district.

In the case of re-interviews, the interviewer was required to question the same person interviewed in 1959 unless deceased or separated from spouse and removed from the community. Sex was randomly assigned to new names on the sample; this sex assignment was binding unless the respondent was deceased or removed from the community.

1961

The 1961 Anglo sample was the same as the new Anglo sample in 1960; in other words, all Anglos who had been interviewed in the new sample in 1960 were to be re-interviewed in 1961. The 1961 Negro sample to be interviewed consisted of a 50 percent random selection of the 1960 Negro sample. The interviewer was required to re-interview the same person interviewed in 1960 unless deceased or separated from spouse and removed from the community.

APPENDIX B. DEMOGRAPHIC AND ECONOMIC CHARACTERISTICS OF COUNTIES AND COMMUNITIES OF ORIGIN FOR MOST MEXICAN-AMERICAN RESPONDENTS

County and Major Community	% Spanish Surname in Total Population in 1960	% Change in Population 1950-1960		Males 14 and Over Unemployed		% Operatives and Above Males 14 and Over	Median Family Income		Median Years of School	
		Spanish Surname in County	Total for City	Non-Spanish Surname	Spanish Surname		Total	Spanish Surname	Total in County, Males 25+	Spanish Surname Persons 25+
La Salle Cotulla	64.2	−24.0	−10.4	4.3	13.0	39.3	$2,296	$1,585	4.4	1.4
Bexar San Antonio	37.4	45.3	43.9	3.5	7.7	76.5	4,766	3,446	10.0	5.7
Zavala Crystal City	74.4	22.8	26.4	3.1	16.8	43.3	2,314	1,732	4.5	2.3
Frio Pearsall	61.8	0.0	10.6	2.6	10.3	50.5	2,676	1,666	6.5	2.3
Webb Laredo	79.9	9.0	16.9	4.9	12.2	66.6	2,952	2,425	6.8	5.4
Maverick Eagle Pass	77.6	24.4	66.2	4.0	20.6	57.3	2,523	2,047	5.7	3.9
Nueces Robstown	38.1	43.2	41.1	5.4	9.9	72.1	4,908	2,974	10.0	4.5

Sources: 1960 U.S. Census of Population: PC(1) 45A, Tables 7, 82; PC(1) 45C, Tables 83, 86, 84; PC(2) 1B, Tables 14, 15.
U.S. Bureau of County and City Data Book, 1967: Appendix A, Table A-2.
Special Report P-E, No. 3C, Table 9, 1950.

APPENDIX C. DEMOGRAPHIC AND ECONOMIC CHARACTERISTICS OF COUNTIES OF ORIGIN FOR MOST NEGRO RESPONDENTS FROM MISSISSIPPI, 1960

County	Percent Nonwhite of Total Population	Percent Unemployed 14 and Over		Percent Operatives and Above, Males 14 and Over		Median Family Income		Median Years of School	
		White	Nonwhite	White	Nonwhite	Total	Nonwhite	Total	Nonwhite
Union	17.5	4.4	7.5	52.8	19.2	$2,274	$1,448	8.8	6.3
Clarke	39.2	5.0	7.3	68.9	28.6	2,343	1,281	8.3	5.5
Lauderdale	34.9	4.6	7.1	87.9	43.9	3,767	1,848	10.1	6.8
Pontotoc	19.2	5.6	5.7	53.5	22.6	1,903	982	8.8	6.6
Holmes	71.9	3.3	7.8	69.6	19.6	1,453	895	7.7	5.8
Lee	25.4	5.3	4.4	75.2	31.5	3,488	1,843	9.6	7.1
Chickasaw	38.7	5.3	6.6	68.0	23.5	2,484	1,052	8.9	7.1

Sources: U.S. Census of Population 1960: PC(1) 26C, Tables 35, 36, 82, 83, 84, 87, 88.

APPENDIX D. POPULATION BY RACE OR ETHNICITY AND LOCATION, 1960

	Total	White Total	Anglo[1]	WPSS[2]	Nonwhite Total	Negro	Other
			White			Nonwhite	
United States	179,323,175	158,831,732	154,870,908	3,960,824	20,491,443	18,871,831	1,619,612
Wisconsin	3,951,777	3,858,903	—	—	92,874	74,546	18,328
Racine (Urban Places)	89,144	84,332	—	—	4,812	4,738	74
5 S.W. States[3]	29,304,012	26,578,041	23,113,042	3,464,999	2,727,971	2,171,444	556,527
Texas	9,579,677	8,374,831	6,957,021	1,417,810	1,204,846	1,187,125	17,721
LaSalle County	5,972	5,965	2,133	3,832	7	7	0
Cotulla	3,960	3,955	1,496	2,459[4]	5	5	0
Mississippi	2,178,141	1,257,546	—	—	920,595	915,743	4,852
Union County	18,904	15,592	—	—	3,312	3,312	0
New Albany	5,151	3,783	—	—	1,368	1,368	0

Sources: U.S. Census of Population 1960: PC(1) 1B, Table 56; PC(1) 26B, Tables 22, 28; PC(1) 45B, Tables 22, 28, PC(1) 51B, Table 21; PC(2) 1B, Tables 2, 15.

1. Anglo population determined by subtracting WPSS from white figures.
2. White persons with Spanish surname. This figure was computed from data in Grebler, Moore and Guzman, **The Mexican-American People.**
3. Arizona, California, Colorado, New Mexico, and Texas.
4. Total number of persons in occupied units, HC(1) No. 45, Table 40.

APPENDIX E. POPULATION BY RACE OR ETHNICITY AND LOCATION, 1960 (IN PERCENTAGES)

	United States	5 S.W. States	Texas			Mississippi			Wisconsin	
			Total	LaSalle County	Cotulla	Total	Union County	New Albany	Total	Racine
White	88.6	90.7	87.4	99.9	99.9	57.7	82.5	73.4	97.6	94.6
Anglo[1]	86.4	78.9	72.6	35.7	37.8[3]	—	—	—	—	—
WPSS[2]	2.2	11.8	14.8	64.2	62.1[3]	—	—	—	—	—
Nonwhite	11.4	9.3	12.6	.1	.1	42.3	17.5	26.6	2.4	5.4
Negro	10.5	7.4	12.4	.1	.1	42.0	17.5	26.6	1.9	5.3
Others	.9	1.9	.2	.0	.0	.2	.0	.0	.5	.1

SOURCES: U.S. Census of Population 1960: PC(1) 1B, Table 56; PC(1) 26B, Tables 22, 28, PC(1) 45B, Tables 22, 28; PC(1) 51B, Table 21; PC(2) 1B, Tables 2, 15.

1. All Anglo percentages are computed by subtracting WPSS from white population figures.
2. White persons with Spanish surname; percent for U.S. computed from data given in Grebler, Moore and Guzman, **The Mexican-American People,** Appendix A, pp. 606-607.
3. Percentage is computed from data obtained in HC(1) No. 45, Table 40—total no. of WPSS in occupied units.

[341]

THE UNIVERSITY OF IOWA

IOWA CITY, IOWA   52240

Department of Sociology and Anthropology

Perhaps you will recall that you were visited in 1959 by one of the interviewers who worked for us on the Racine community project. Although we attempted to reach you and interview you in 1960, we were unable to do so.

During the summer of 1970 we hope to interview everyone again whether they still live in Racine or not. For this reason, we are writing to you now in order to find out where you are, and how things are going. Many changes have taken place in Racine since 1959 and people have had all kinds of experiences since then. We hope to find out a little bit more about you now and much more when we visit with you in 1970.

Enclosed you will find a short set of questions about you and your family. We would very much appreciate it if you would write out the answers to these questions and return them to us as soon as possible. If you move to another home in Racine between now and the summer of 1970, we would like to have you send us the enclosed card telling us your new address. If you move out of Racine to another community, we would also like to have you keep in touch with us by returning the card. If you save this letter and the card, that will help you remember us.

We would like to again thank you for the assistance that you gave to us by talking with our interviewer in 1959. The study that we made was a success only because people like you were willing to spend some time with our interviewers. We shall not forget you and are looking forward to hearing from you and then seeing you again in 1970.

Sincerely yours,

Lyle W. Shannon

LWS/jm
Enc.

APPENDIX F.

NAME INDEX

SUBJECT INDEX

ABOUT THE AUTHORS

LYLE W. SHANNON received the B.A. from Cornell College in Mount Vernon, Iowa, in 1942, and earned his Ph.D. in sociology at the University of Washington in Seattle in 1951. He taught at the University of Wisconsin for ten years and in 1962 accepted the position of Professor of Sociology and the chairmanship of the Departments of Sociology and Anthropology at the University of Iowa. He was chairman of the joint department until 1970. Since that time he has been director of the Iowa Urban Community Research Center and Professor of Sociology at the University of Iowa.

MAGDALINE SHANNON, presently a graduate student in the Department of History at the University of Iowa, received her B.A. from the College of St. Benedict in St. Joseph, Minnesota, and previously taught in an all-black high school in Wisconsin.